MICROCOMPUTERS AND LABORATORY INSTRUMENTATION

SECOND EDITION

MICROCOMPUTERS AND LABORATORY INSTRUMENTATION

SECOND EDITION

David J. Malcolme-Lawes
King's College London
London, England

PLENUM PRESS • NEW YORK AND LONDON

Library of Congress Cataloging in Publication Data

Malcolme-Lawes, D. J.
  Microcomputers and laboratory instrumentation.

  Bibliography: p.
  Includes indexes.
   1. Physical instruments—Data processing. 2. Physical laboratories—Data process-
ing. 3. Microcomputers. I. Title.
QC53.M27 1988   530′.7′0285416   88-5794
ISBN 0-306-42903-9

© 1988 Plenum Press, New York
A Division of Plenum Publishing Corporation
233 Spring Street, New York, N.Y. 10013

Printed in the United States of America

To Meryl

PREFACE

The invention of the microcomputer in the mid-1970s and its subsequent low-cost proliferation has opened up a new world for the laboratory scientist. Tedious data collection can now be automated relatively cheaply and with an enormous increase in reliability. New techniques of measurement are accessible with the "intelligent" instrumentation made possible by these programmable devices, and the ease of use of even standard measurement techniques may be improved by the data processing capabilities of the humblest micro. The latest items of commercial laboratory instrumentation are invariably "computer controlled", although this is more likely to mean that a microprocessor is involved than that a versatile microcomputer is provided along with the instrument.

It is clear that all scientists of the future will need some knowledge of computers, if only to aid them in mastering the button pushing associated with gleaming new instruments. However, to be able to exploit this newly accessible computing power to the full the practising laboratory scientist must gain sufficient understanding to utilise the communication channels between apparatus on the laboratory bench and program within the computer.

This book attempts to provide an introduction to those communication channels in a manner which is understandable for scientists who do not specialise in electronics or computers. The contents are based on courses given to undergraduate and postgraduate science students at King's College London and to numerous industrial and government scientists who attended short courses at the College. The objective of those courses was to provide scientists with an understanding of how modern microcomputers can communicate with laboratory apparatus for measurement and control purposes. It was not expected that all the students would have to design and build interfaces to achieve their ends, but rather that they should understand the principles on which interfaces operate and the capabilities and limitations of practical devices, so that they could design experiments in their own fields with a foundation knowledge of how a microcomputer could be employed. The courses were closely associated with practical experience gained on microcomputers and a variety of items of standard laboratory instrumentation. Of course that element is not included in the present text, but the fact remains that this book is intended to be of assistance to the practical scientist.

While not designed as a do-it-yourself guide to building particular electronic circuits, a number of interfacing circuits are discussed in some detail and the readers may well be able to develop these to suit their own needs. The circuits described are derived from the author's own experience, which is limited to systems associated with PET, Commodore 64, Amiga, Apple, BBC, Spectrum and IBM PC microcomputers. The devices available for use in signal handling continue to increase rapidly. Even over the last year a number of new devices have appeared (particularly newer ADC devices and LSI circuits) that would be useful for a number of the tasks discussed. However, the principles of communication between microcomputers and laboratory systems are not changing quite so rapidly, and it is hoped that the examples will be found useful.

I wish to express my thanks to Drs. Kevin Jones and Philip Moss of King's College London for their helpful advice and criticism during the preparation of this manuscript, to the research students who have suffered changes in the apparatus while circuits were tested, and to Dr Celio Pasquini who has ably assisted in the design of many of the circuits used with IBM PCs, who read the manuscript and thus helped me avoid too many errors. I would also like to thank my family for their patience during the long period of preparation which was the cost of this second edition.

Finally I wish to thank the Royal Society for their support over the last five years. It is only through the support of the Royal Society that scientists in the UK are able to take time from the normal torrent of life to turn to a new field of study, and for the opportunity so provided I will always be grateful.

King's College London   David J. Malcolme-Lawes
December 1987

CONTENTS

CHAPTER 1

INTRODUCTION

One of the consequences of a major change in technology is that a large number of familiar ideas, procedures and equipment quickly become replaced by less familiar devices and ways of doing things. The world embarked on one such revolutionary upheaval about a decade ago and, although it will be many years before the full impact of computers is felt, there is no doubt that we are already feeling the effects of their presence and should ensure that we are in a position to take full advantage of the even more dramatic changes which still lie ahead. The scale of the technological changes to come will undoubtedly be enormous, just as was the case during the first industrial revolution, and go well beyond the perceptions of those who think that "it" has already happened. These changes will eventually affect almost every aspect of our lives, from education to leisure, from housekeeping to heavy industry. In this book we are concerned with the beginings of this change in one small area of activity - the laboratory, and in particular with the electronic instrumentation used in the laboratory for measurement and control purposes.

Let us start by making quite clear the aim of this book. The objective is to provide scientists who do not specialise in electronics or computers with an introduction to the basic aspects of the use of common, mass produced microcomputers for communicating with and controlling experimental equipment. It is not part of the objective to discuss robotics, the mathematical procedures of data handling or sophisticated electronic signal processing techniques. Neither is the use of microprocessors in laboratory instrumentation to be discussed and it is not assumed that the reader has any special knowledge of these devices - other than that a microprocessor is one of the component parts of a microcomputer. In this chapter we discuss briefly some general aspects of microcomputers and their relation to laboratory instrumentation, and outline the level of knowledge required to appreciate subsequent chapters. In chapter 2 we examine the basics of the electrical signals commonly encountered in laboratory systems, and in chapters 3 and 4 we discuss the major elements of the electronic circuits used for handling analog and digital signals respectively. In chapter 5 we look at the modern microcomputer, some of its peripherals which are of value in the laboratory, and a few of the details of its internal organisation and function - where these are related to communication with external equipment.

One particular kind of microcomputer (the PC - a name used in this book to refer to microcomputers designed to be compatible with the IBM PC, PC/XT or PC/AT models) and its related software have come to assume such a major importance for low-cost laboratory applications that chapter 6 is devoted to a somewhat more detailed discussion of the characteristics of the PCs. Chapters 7, 8 and 9 examine the major techniques which can be used for communication between a microcomputer and other items of analog or digital signal handling equipment - a subject known almost universally as interfacing. In chapter 10 we consider how to approach the problem of designing an instrumentation system which uses a microcomputer as its control centre, looking at both the hardware and software aspects of the problem and concluding with a case study based on an instrument designed at King's.

It must be pointed out that the subject matter of the book is intended to overlap with a number of traditional fields, and that the topics discussed are covered solely for their relevance to the application of microcomputers in practical laboratories. The coverage can undoubtedly be criticised for its omission of topics necessary for a wider understanding of electronics or computers, and could equally be regarded as unnecessary for the man who just wants to connect a computer to an instrument to collect some data. However, the nature of laboratories is changing as a result of the technological reorientation that computers are bringing, and in this author's view many scientists, whatever their specialist field, do require a quantity of knowledge in these areas.

1.1 Laboratory instrumentation and microcomputers

Laboratory instrumentation assists the scientist by enabling him to make measurements of a wide range of physical, chemical and biological parameters, and by automating the control of measurements, processes and recording functions. Examples of just a few of the measurement and control functions of widely used items of laboratory instrumentation are listed in table 1.1.

Table 1.1 Examples of measurement and control
functions of commonly used laboratory instruments

Instrument	Measured property
thermometer	temperature
manometer	pressure
photometer	light intensity
pH meter	hydrogen ion activity
GM counter	radioactivity
multimeter	voltage, current, resistance
clock	time

Instrument	Controlled property
thermostat	temperature
manostat	pressure
potentiostat	applied voltage
monochromator	wavelength transmitted
timer	time interval
flostat	flow rate

Table 1.2 Examples of low cost microcomputers

Models	Manufacturer
Amiga 500 & 2000	Commodore
Amstrad 1640	Amstrad
Apple II	Apple Computers
Archimedes	Acorn Computers
Atari 520 & 1040	Atari
BBC Master Series	Acorn Computers
Commodore 64 & 128	Commodore
Compaq Deskpro 386	Compaq Computers
HP125, HP300	Hewlett Packard
IBM PC, XT & AT	IBM
Macintosh	Apple Computers
Nimbus	Research Machines
TRS80	Tandy

Modern instrumentation allows sophisticated combinations of many of the basic measurement and control functions to provide systems capable of the direct and highly automated measurement of complex quantities. Examples include systems such as thermal luminescent dosimeters, automatic liquid scintillation counters, materials testing equipment, magnetic resonance spectrometers, infrared and ultraviolet absorption spectrometers and gradient elution chromatographs. However, although well established instrumental techniques have benefitted from recent developments in electronics, the situation for laboratories with limited resources and for research groups working in new areas of measurement and control is less favourable if commercial systems which meet their specifications are not available.

The dramatic growth in the availability of low cost microcomputers over the last few years has initiated significant changes in the development of laboratory instrumentation, not only by allowing simplification of the operation of many common instruments, but also by providing the means for the development of new instruments based on methods of measurement and control complexity which were relatively inaccessible to the previous generation, such as real-time Fourier transform techniques and diode array detectors for spectroscopy. Some of the most popular microcomputers, available at costs ranging from less than $100 to over $5000, are listed in table 1.2. Any of these micros may be used for the moderately high speed measurement of signals, storing of measurements and data, controlling electrical and mechanical equipment, performing arithmetical and logical manipulations, and the displaying or recording of information. Naturally some functions are more easily accomplished with the improved convenience of widely available hardware and software components using the higher priced models, but there are many less demanding functions for which a relatively inexpensive micro is not only quite adequate but also more appropriate. Thus there is little point in using a Compaq 386 to act as a chromatographic chart recorder.

Table 1.3 Typical BASIC language instructions

1	:REM Remarks for ease of reading
10 INPUT X	:REM input data and store in X
20 Y = 2+X/3	:REM calculate Y from expression
30 GOTO 100	:REM control transferred to 100
40 PRINT "Y= "; Y	:REM output message & Y
50 IF X=0 THEN GOTO 100 :REM goto 100 if X is 0	
70 STOP	:REM stop execution of program
60 FOR I=1 TO 100	:REM instructions between FOR and
- - - - -	:REM NEXT repeated for values of
69 NEXT I	:REM I from 1 to 100 inclusive
80 GOSUB 5000	:REM control transferred to line
5000 Y=0	:REM 5000 then instructions obeyed
5999 RETURN	:REM until RETURN is encountered

Microcomputers offer the laboratory scientist a new dimension in instrumentation because they can be programmed to perform the tasks of complex electronic circuits, and the program can be modified until it does precisely what is wanted. Furthermore, when properly designed and tested, microcomputer based systems can offer fast and reliable operation over long periods of time and yet can be rapidly reconfigured to perform a totally different function when necessary. They offer the laboratory scientist the ability to control instrumentation during long term unattended operation, to automate routine measurement functions for large numbers of samples, to record and process large amounts of data, and to present the result of a measurement based on many or different quantities.

The value of the modern microcomputer lies in the fact that it can be programmed. Almost anyone with a secondary school education can master the elements of programming a microcomputer in an afternoon using the most common language (BASIC). The principal instructions of the BASIC language are listed in table 1.3, although we are not going to cover the language further or discuss its normal use here. Indeed we will assume that the reader is familiar with both the appearance of microcomputers and with the elements of BASIC (or any other BASIC-like language, such as FORTRAN) when we come to discuss the operation of micros and communication with them. (There is a large number of books available on the use of BASIC and it is to be highly recommended that the reader who is not already familiar with the language should study one of these before going beyond chapter 4 in this text.)

1.2 Measurement systems

Let us begin by examining the nature of a conventional laboratory measurement system as illustrated by the block diagram shown in fig 1.1. The system consists essentially of three components. The first is the physical or chemical process which is the basis of the

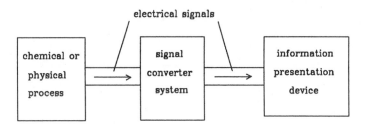

Fig 1.1 *The basic components of a conventional measurement system.*

measurement, and this can be almost anything from a mouse taking a piece of cheese, to carbon-13 atoms absorbing radiofrequency energy at a specified frequency in the magnetic field of an NMR system. The last component is some kind of information presentation or display device which presents a feature of the quantity being measured in a form which can be noted by a human operator. This may be a simple chart recorder trace or a display showing a numerical value. Generally in between there is a second component which acts as a signal converter, converting the signal produced by the first component into the signal required by the third. In most cases the signals are of an electrical nature, and the signal converter is an electronic circuit.

As an example of a simple conventional measurement system consider the basic pH meter illustrated in fig 1.2. In this case the first component of the system is the pH electrode in contact with the solution being tested. The third component is a digital display, allowing readings of pH to be observed directly to two decimal places (eg. 7.04). The signal converter component in this case is an electronic circuit which converts the electrical signal generated by the pH electrode into the whatever signals are required to make the digits 7.04 appear on the display.

Let us now consider a computer based laboratory measurement system, using the simple example illustrated in fig 1.3. In this example the first component may be the same as the first component of a conventional measurement system, thus it might be the pH electrode as

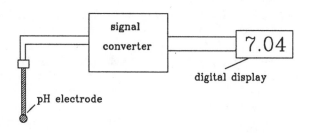

Fig 1.2 *The basic components of a typical pH measurement system, incorporating an electrode, a signal converter and a display.*

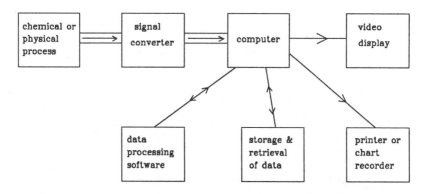

Fig 1.3 *The elements of a computer based measurement system,*
including both hardware and software components.

used above. However, instead of just an information display device we now have a computer
receiving the signals generated by the measurement process. We still require a signal
converter system, to convert the signal produced by the first component into the types of
signals which can be understood by the computer, and this will be a different kind of
electronic circuit from that used in the conventional measurement system.

Now, using the computer in a most elementary role we could program the computer (ie.
use "software") to produce signals suitable for an information display, which may be just a
display of numbers on a video screen, a fairly conventional chart record, or even an elaborate
multicoloured diagram annotated with helpful details (such as a spectrum with peaks
highlighted and axes labelled). In that case the computer based measurement system would
provide a similar result to that obtainable from the conventional measurement system. For
example, if the first component had been the pH electrode, the computer could be
programmed to display the pH on a video screen and we would have an expensive pH meter.

However, the computer based system has a number of advantages to offer. Firstly a
computer can be programmed to process data before displaying a result. Suppose that we use
our computerised pH meter for a system in which a slow chemical reaction is occurring in the
sample at a rate proportional to the concentration of a substance X and that one of the
reaction products causes a change in the pH of the sample as the reaction proceeds. The
computer could be programmed to monitor the pH variation over a period of time, calculate
the rate of the reaction, and display the concentration of X. This is surely a big improvement
over manually recording a series of pH readings and then sitting down with a calculator to
calculate the quantity of interest. Secondly a computer may be programmed to store results
for later use, or to compare a result with one obtained earlier. Thus a spectrum recorded on a
computer may be filed away on, say, disk and at a later time retrieved and overlaid with the
spectrum of another sample, to aid perhaps in the identification of a sample. Thirdly the
computer is a flexible device, its software may be changed at will. So our pH meter may be
converted into a chloride ion monitor by changing the electrode to a chloride ion selective

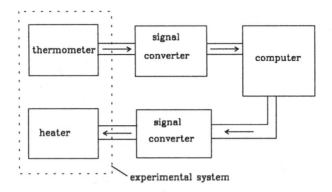

Fig 1.4 *A computer based system with control and measurement functions.*

electrode and the computer's program so that it displays the chloride ion concentration. Similarly a single computer may be connected to several different measurement devices and with a suitable program may be used to monitor several different properties at once, such as pH, chloride concentration, temperature and uv absorbance at 280nm.

Finally a computer based measuring system may also be programmed to effect and control the conditions in the system on which measurements are being made. For example, fig 1.4 shows the computer producing electrical control signals which are then used by a device (say, a heater) to effect the experimental system in a specified manner (eg. by changing its temperature), and monitoring electrical signals produced by a measuring device (say, a thermometer), so that the computer is able to keep track of the effects of its control signals.

Each of these aspects of a computer based measuring system has been dependent on the hardware (ie. the electrical circuits) only to the extent that signal converters are required to translate the signals used or produced by various components into those which can be understood by the computer. But the versatility and power of the the system really comes from our ability to use software (ie. the computer's program) to manipulate these signals in an infinitely extendable variety of ways. This book is largely about the hardware aspects involved in connecting computers to other components. Although there will undoubtedly be developments in this field they will probably fall in the category of technical improvements rather than revolutionary changes. We discuss software in only a limited way, and from the much more restricting philosophy that major changes in this area still lie ahead. However, the reader should always keep in mind that any computer based system remains no more than a collection of low-value components until the software is created to interpret the input signals and generate the outputs.

1.3 Electronic black boxes

Throughout this book we will be discussing electronic circuits which form the basis for the hardware connection between a microcomputer and any external devices. In chapter 2 we

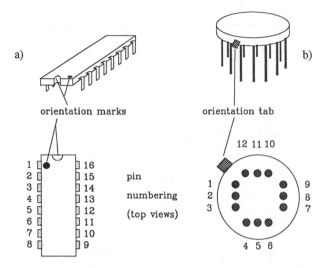

Fig 1.5 *Two of the commonly encountered packages used for integrated circuits.
a) the DIL package, and b) the TO-8 can.*

examine the nature of the electrical signals most commonly encountered in the laboratory and discuss some important aspects of signal transmission. In chapters 3 & 4 we discuss the types of electronic circuits relevant to signal converter applications. In all cases it is assumed that the reader has a basic knowledge of electrical components and electricity, covering Ohm's law and elementary ac theory, although a detailed knowledge of electronics is not required. (The symbol R will be used to indicate resistance values in ohms, k for kilohms and M for megohms.) Our discussion will be almost entirely confined to integrated circuits (ICs), which can be regarded as "black box" electronic devices with precisely defined properties (ie. applying particular signals at certain connectors of the IC results in predictable signals appearing at other connectors).

Most of the ICs mentioned are readily available from a variety of manufacturers and suppliers, although the complete code numbers used to identify particular circuits may vary with the manufacturer. For example, the 709 operational amplifier (see chapter 3) may be identified by the codes SN72709N, MC1709G, LM709C or a variety of other codes. Where a device is available from many sources, circuit diagrams containing the device are labelled using only the device code (eg. 709) and not with the manufacturer or packaging codes. Where any ambiguity may result the code of the commercial quality device offered by National Semiconductor Corporation has been used (generally these begin with LH, LF or LM depending on whether the device is of hybrid, BIFET or monolithic construction respectively).

Many of the devices discussed are available in a variety of packages. Nearly all of the circuits shown in this book have been used in the author's laboratory, where the circuits were constructed using commercial quality plastic packages (these are usually black) of the dual-in-line (DIL) configuration illustrated in fig 1.5a. DIL packages have two rows of connection pins, each pin being 0.1 inch from its nearest neighbour and the two rows being 0.3 or 0.6 inches apart, and the number of pins varying from 4 to 40. At no time has the author found it necessary to use the more expensive ceramic DIL packages where the plastic packaged devices have been available (some importers only stock the ceramic packaged devices), nor has the author deliberately purchased any devices specified to higher than commercial quality (military and certain other quality devices are specified over wider operating temperature ranges and may have a tighter spread of characteristics than their commercial equivalents). A variety of other packages are available for many analog devices, the 12 pin TO-8 can shown in fig 1.5b being one of the most commonly encountered. Some of the devices discussed are only available in this form.

1.4 A practical footnote

This book was not conceived as a "do-it-yourself" guide to building computer interface systems, although hopefully some readers may find that the subject is not as complex as they may have thought and so be encouraged to do just that. For such readers the following brief practical comments may be of assistance. The majority of our circuits have been constructed on pre-drilled IC circuit boards (we use Eurocards) using "wire wrapping", a technique by which wires are wrapped without soldering around the pins of IC sockets to convey signals from one part of a circuit to another. This technique is to be highly recommended as only a couple of inexpensive tools are required and wiring errors can be easily rectified. Unless a considerable amount of test gear and expertise is available the "insulation displacement" form of wire wrapping (in which plastic covered wire is wrapped around the socket pins and the sharp edges of the pins cuts through the plastic to make contact with the wire) or wire insertion is not advised. The time wasted in finding a faulty joint can more than offset the time spent in conventional wire wrapping (where the insulation is removed from the wire before wrapping). For high speed (>10 MHz) and fast TTL (see chapter 4) we have preferred to rely on printed circuit boards.

Many of the systems described involve multiwire connections between a microcomputer and an auxiliary electronic circuit such as an interface system. For such connections the author is a convinced advocate of "insulation displacement" connection systems (available at relatively low cost from most electronic hardware supply houses). In these the connection between the conductor wire and the connector pin is made by teeth (tines) on the connector which are forced through the plastic insulation around the wire, usually with the aid of a simple bench vice. The use of this technique almost forces a high degree of neatness as special soft plastic covered ribbon cables have to be used.

Most microcomputers are powered from the mains, so don't poke around inside with screwdrivers and bits of wire while the computer is still plugged in - the repair costs could be as much as a new computer. Many of the cheaper micros are low voltage devices powered from separate transformer/rectifier units. In either case be careful about taking power from the micro to drive external circuits like interfaces; some manufacturers have provided power supplies which are barely adequate to run the computer let alone any extra circuits. Although the PC is an exception (it was obviously designed to have extra circuits plugged into its IO slots), with most micros it is usually wise to provide separate power supplies for interfaces using more than half a dozen TTL ICs or equivalent, and in such cases it is particularly important to ensure that the micro does not produce interference (see chapter 2) in any analog interface circuits. TVs and video monitors tend to radiate a fair amount of high frequency signal. The 0 V connections of the micro and the externally powered devices will need to be connected together but some care should be taken about whether the micro can be grounded - certainly check for any voltage between the micro's 0 V level and mains ground first, and if it is more than a few millivolts be careful about connecting to ground. If a separate ground connection for the micro has to be provided then this should be done at the power supply circuit and not from a point in the middle of the logic circuits. And always be aware that dabbling inside the computer generally invalidates any warranty it had.

CHAPTER 2

THE BASICS OF LABORATORY SIGNALS

Parameters measured in laboratories cover a wide range of physical properties and effects, such as luminescence, electromagnetic energy absorption (uv, ir, nmr, etc.), electrochemical potential, temperature, pressure, radioactivity, etc., in addition to simple coordinate changes, such as distance, angle, velocity or acceleration. However, virtually all such physical characteristics may be converted into some form of electrical signal by an appropriate energy conversion device known as a transducer. Some typical transducers are listed in table 2.1, along with the nature of the particular energy stimulus for which each is designed.

2.1 Transducers

Transducers fall into one of two primary categories: charge generating transducers and impedance transducers. The common feature of charge generating transducers is illustrated in fig 2.1. A circuit element known as an electrode gathers charge liberated by some physical process and so becomes positively or negatively charged. Unless we are concerned with electrostatic measurements, a second pathway is required to return charge to the environment of the transducer. This is usually achieved through a "ground" or "earth" connection at a second electrode, although it is possible to "float" systems at potentials other than ground.

Table 2.1   Examples of laboratory transducers

Transducer	Energy stimulus
Coil	changing magnetic field
Electrochemical electrode	chemical potential
Ionisation chamber	ionising radiation
Microphone	sound/vibration
Photomultiplier	uv/visible photons
Photodiode	light
Piezo-electric crystal	pressure
Resistance thermometer	temperature
Strain gauge	deformation
Thermocouple	temperature, ir

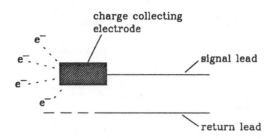

charge collecting
electrode

e⁻

e⁻

e⁻

e⁻

signal lead

return lead

Fig 2.1 *The principle of a charge generating transducer, in which an electrode collects*
charge (in this case electrons) liberated by some physical process.

The effect of this gathered charge is determined in part by the electrical impedance of any
measuring circuit connected to the electrodes' signal and return leads, and we shall consider
this aspect of the measurement process in section 2.3. For the present we merely observe that
the electrons collected at the electrode may be allowed to flow along the signal lead, in which
case they constitute an electric current and the transducer provides a source of current.
Alternatively the electrons may be prevented from flowing along the signal lead - by the
presence of a very high impedance to current flow between the signal and return leads - in
which case the potential of the electrode is changed by the process of collecting the charge.
The voltage change produced on the electrode (or more precisely the difference in potential of
the electrode before and after charge collection) may be calculated using the equation

$$\Delta V = Q/C_T \qquad\qquad\qquad (2.1)$$

where   Q is the charge collected,
         C_T is the capacity of the transducer electrode, and
         ΔV is the voltage change produced.

Thus a transducer of this class may act as a source of current or voltage, depending on the
technique selected for the measurement of the charge collected. Examples of this first class of
transducer include devices such as photomultiplier tubes, ionisation chambers and
semiconductor detectors for nuclear particles.

The second major class of transducer is the impedance transducer, in which the electrical
impedance of the device is influenced either by its physical deformation, such as the
stretching of a resistive wire, or by some property of its surroundings, such as the
temperature, pressure, light level or magnetic field. In this case the impedance of the
transducer is monitored using a circuit illustrated in its simplest form in fig 2.2a. A signal is
applied to a series impedance network made up of the transducer and a second impedance of
fixed or adjustable characteristics known as a reference impedance. The potential at the
junction of the two impedances is measured and provides information about the variation of
the transducer impedance relative to the reference impedance. As the reference impedance
may in practice be the input impedance of a current measuring circuit, this class of transducer
may also be considered to act as either a source of current or a voltage source. Typical

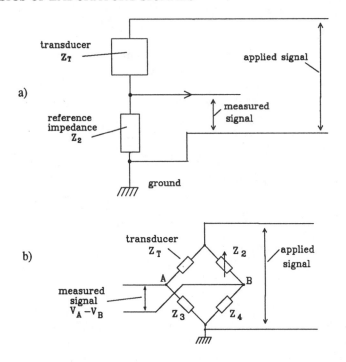

Fig 2.2 *Common arrangements for measurements using an impedance transducer which undergoes an impedance change as a result of a physical stimulus. (a) A simple circuit for measurement of impedance relative to a reference impedance. (b) A bridge circuit for the measurement of small changes in the transducer's impedance.*

transducers of this class include resistance thermometers, strain gauges, magnetometer coils, light sensitive resistors and diodes, and capacitance devices for the measurement of distance or dielectric constant.

When an impedance transducer is being used for the measurement of relatively small changes in a physical characteristic, a bridge circuit is often used so that the "background" level of the measured signal may be effectively cancelled out or "nulled". A generalised bridge circuit is shown in fig 2.2b. The variable impedance, Z_2, may be adjusted so that

$$Z_T / Z_3 = Z_2 / Z_4$$

in which case the bridge is balanced, the voltages at points A and B are identical, and the measured voltage difference signal, $(V_A - V_B)$ is zero. Any change in the transducer's impedance will now unbalance the bridge and produce a difference signal related to the impedance change. Bridge circuits are particularly simple to implement when the transducer's impedance is purely resistive, as the impedances Z_2 - Z_4 may be resistances and the measured and applied signals may be dc voltage levels. Such arrangements are commonly encountered in measurement systems based on strain gauges.

2.2 Measurement signals

We have seen that transducers act as the sources of current or voltage, and of course to these may be added signal sources for which the property being measured is of an electrical nature (eg. electrochemical cell potentials, conductivities, etc.). However, the signals likely to be met in a laboratory may also have an important time dependence, either because the effect responsible for the transducer output exhibits a time dependence (eg. a vibration or a repeating or decaying signal), or because the signal may be varied in some way to assist in the measurement (eg. the modulation of the magnetic field in magnetic resonance spectroscopy). The types of signals most commonly encountered in laboratory instrumentation are summarised in fig 2.3.

Figure 2.3a illustrates a constant (or slowly varying) voltage or current which has a time average value which is non-zero. This kind of signal is called a direct current, dc, signal, even when one is talking about a voltage or a signal which does vary with time. A dc signal is produced, for example, by a transducer monitoring temperature or pressure in a static system. In this case the quantity of importance is the magnitude of the signal, and the signal may be electronically filtered to remove short term variations, such as the effects of noise or interference (see section 2.4). In fig 2.3b a representation of an oscillating or alternating (ac) current or voltage is shown, and the principle feature of this type of signal is that its mean value is zero. Signals of this nature may be found in a radiofrequency receiver or an ac bridge circuit for the measurement of the conductivity of a solution. Probably the most common ac signals are those which have a sinusoidal time dependence, eg.

$$V = V_0 \sin(2\pi ft) \tag{2.2}$$

where V is the signal voltage at time t,
       V_0 is the peak amplitude and
       f is the normal frequency of the oscillation in hertz.

In this case either the absolute magnitude or the frequency of the signal (or both) may be the important parameter for measurement, and filtration of the signal to remove variations of a different frequency may be accomplished easily only if the frequency range over which the signal may vary is limited. The magnitude of an ac signal may be expressed in terms of its peak values (V_0) or its root mean square (RMS) value, which for a sinusoidal signal is about $0.707V_0$.

Figure 2.3c illustrates a pulse signal. A current pulse of this kind is produced by a photomultiplier tube detecting brief flashes of light or single photons. In this case filtration of the signal is virtually impossible in fast systems and alternative techniques have to be used to minimise the effects of noise and interference. The important parameter associated with pulse signals may be the magnitude of the pulse (which may be its height or its integrated area) or the number of pulses per second. It is the difficulty in measuring the height of very narrow pulses which often leads to the conversion of such pulses into the "long tail pulse" form of signal illustrated in fig 2.3d. Generally the tail of the signal has a characteristic mathematical form, such as an exponential decay. The conversion may be carried out by the integration of

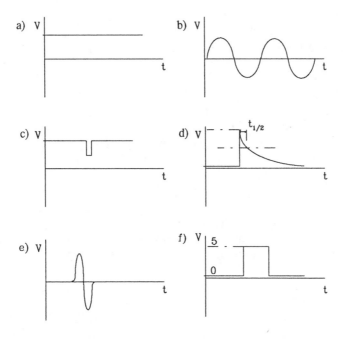

Fig 2.3 *Types of signals commonly encountered in laboratory instrumentation. In each case the variable signal may be a current or a voltage. a) a dc signal, b) an ac signal, c) a unipolar pulse, d) a decaying signal or a long tail pulse, e) a bipolar pulse, and f) a logic pulse.*

fast pulses, as discussed in section 3.9, although the same form of signal may also be characteristic of measured physical properties - such as the intensity of a laser flash, or the number of bacteria surviving after a chemical has been added to their growth medium.

The long tail signal is produced as a voltage pulse from a charge sensitive transducer (see eqn (2.1)) when the charge is collected rapidly, but leaks away from the collecting electrode more slowly (usually through a high resistance). Examples include photomultipliers detecting scintillations and X-ray and nuclear particle detectors. For this signal the important parameter may be the pulse height or the time of decay of the signal, which is usually taken to be the time interval between the time at the peak of the signal and the time at which the signal falls to one half (or sometimes 1/e) of its peak value. While it can be illustrated in a single diagram, this type of signal actually covers a wider range of measurement techniques than most of the others, simply because of the very wide range of time scales which may be involved. For example, measurements can be made on the half-life of a radioisotope, which may be minutes or hours, or on the lifetime of a fluorescent species, which may be a few nanoseconds. As a result this kind of signal may be the most difficult to measure accurately, particularly as its measurement relies upon maintaining the fidelity of the transducer output through many stages of electronic signal processing, which is difficult to achieve over a wide frequency range.

The pulse signal shown in fig 2.3c is actually a unipolar pulse, as the signal value is always on only one side of the background voltage or current (which may, of course, be zero). This kind of pulse shares a number of characteristics with a dc signal, in spite of the fact that it may involve very fast changes of voltage or current. An alternative type of pulse is the bipolar pulse, shown in fig 2.3e, which has a time-averaged signal level of zero and so may be regarded as a kind of ac signal. Bipolar pulses, in common with other ac signals, may be passed through capacitors which block dc signals and, as we shall see, this "ac coupling" of signals can offer certain advantages. Unipolar pulses have to be treated more carefully: attempting to pass them through a capacitor, for example, can result in their conversion into bipolar pulses and hence a change in the time-averaged signal level.

All of the above signals are examples of analog signals, because their magnitude carries information which can represent the value of a continuously variable quantity. However, an increasingly common type of signal associated with laboratory instrumentation is the logic level signal. An example is provided by the logic pulse shown in fig 2.3f. A logic pulse consists of a transition from one specified logic level to another and back again. In most cases the signal levels involved are 0 V and +5 V, although 0 V and +10 V are also common. (Logic level signals will be dealt with in more detail in chapter 4.) In logic pulse systems the important parameter may be the presence or absence of a pulse, the number of pulses per second or the width of each individual pulse (and the latter may be used in an analog sense). While the number of transducers which generate logic pulses directly is limited (eg. to mechanical devices and some trigger circuits), many pulse producing transducers can be fitted with microcircuit converters which convert current or voltage pulses above a certain threshold into logic pulses (as will be illustrated in fig 4.16). Photon counting photomultiplier systems and channeltron multipliers are ideal candidates for this treatment because the logic pulse signals are virtually immune from noise or interference and may be transmitted over long distances without loss.

2.3 The transducer connection

The electrical signal produced by a transducer is of value only if the signal characteristic of interest can be measured. To make any kind of electrical measurement requires the connection of the transducer's output to the input of a signal converter, a device which converts the transducer's electrical signal into a signal which may be more easily presented to and understood by man. The signal converter's output may drive a display device which may take any convenient form, from a moving needle meter to a picture (provided, say, by an oscilloscope trace) or a sophisticated computer display. The transducer connection is illustrated schematically in fig 2.4, where two wires, known as signal transmission lines, carry the transducer output to the signal converter input, which acts as a load for the signal's energy. It is the signal on these transmission lines which forms the basis of all subsequent measurements, so that, in spite of the fact that these lines are all too frequently neglected in laboratory work, we will consider them in some detail. Much of the discussion also applies to signal transmission lines connecting one stage of a converter system to the next.

For most laboratory signals the transmission system may be regarded as consisting of the elements shown in fig 2.5. The origin of the signal is a two terminal signal source with a

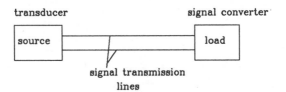

transducer signal converter

source load

signal transmission
lines

Fig 2.4 *The connection of a transducer to a signal converter. In many
cases one of the transmission lines may be grounded.*

source series impedance which is an inherent feature of the source design. The source may
also be associated with a parallel impedance (usually capacitive), although this is generally of
far less importance unless very high signal frequencies are used. Where the signal source is a
transducer, the source series impedance is generally outside the control of the user, and may
range from a low value (say, 10R for a hot filament or 100R for an ir thermocouple) to a very
high value (say, 10^9R for a pH electrode or 10^8R for an illuminated photomultiplier tube).
The signal source and its associated series impedance form the termination at one end of the
transmission line. At the other end is the load formed by the input circuit of the signal
converter, which may be regarded as a measuring device of infinite impedance in parallel
with a parallel load impedance. As the signal pathway is from the signal source to the signal
converter, the series source impedance is often referred to as the output impedance of the
source, and the parallel input impedance of the converter as simply the input impedance of
the converter. The user does have control of the converter's input impedance to the extent
that he can select the most suitable converter input circuit for the system in question.

An important feature of a transmission line terminated at each end by devices with
impedance is the efficiency with which power is transferred from one end of the line to the
other. We first consider the amount of power generated by a simple source which is
producing a constant dc voltage, as illustrated using the equivalent circuit shown
schematically in fig 2.6. (This enables the discussion to be carried out in terms of resistances,
rather than complex impedances.) If the signal source is generating a dc voltage, V_s, and its
series resistance is R_s, then the maximum power generated by the source will be produced
when the maximum current is drawn from it, and this will occur when the output leads are
shorted together. Designating this short circuit current I_s, then the maximum power

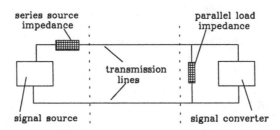

series source parallel load
impedance impedance

transmission
lines

signal source signal converter

Fig 2.5 *The principal features of the general circuit formed by the connection of a signal
source (such as a transducer) to a load (such as a signal converter).*

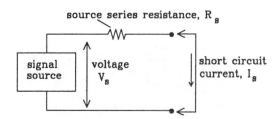

Fig 2.6 *An equivalent circuit for a voltage signal source with its series source resistance,*
shown with the output terminals shorted so that the maximum current may flow.

generation is

$$P_{max} = V_s * I_s = I_s^2 * R_s = V_s^2 / R_s \qquad (2.3)$$

If the output is now connected to a load resistance, R_L (which may represent the input
resistance of a converter circuit), as illustrated in fig 2.7, and if the current flow is now I, then
the power dissipated in the source resistance becomes

$$P_S = I^2 * R_S$$

and that dissipated in the load resistance R_L becomes

$$P_L = I^2 * R_L$$

The total power being dissipated is now

$$P_T = P_S + P_L = I^2 * (R_L + R_S) = V_s^2 / (R_L + R_S)$$

It follows that the power transferred to the resistance R_L is

$$P_L = (V_s^2 * R_L) / (R_L + R_S)^2 \qquad (2.4)$$

which, when combined with eqn (2.3), gives

$$P_L = P_{max} R_L R_S / (R_L + R_S)^2 \qquad (2.5)$$

If $R_L >> R_S$ then the power transferred to R_L approximates to

$$P_L = P_{max} * R_S / R_L$$

Thus if R_S = 50R and R_L = 1M then $P_L = 5*10^{-4}*P_{max}$. Similarly if $R_L << R_S$ then

$$P_L = P_{max} * R_L / R_S$$

and if R_S = 1M and R_L = 50R then $P_L = 5*10^{-4}*P_{max}$.

In both cases very little of the power being generated by the signal source is actually reaching the load. To find the maximum power transferable to R_L for a fixed value of R_s it is necessary to find the value of P_L when the partial derivative, dP_L/dR_L, is zero. Differentiating eqn (2.5) we find

$$dP_L/dR_L = P_{max}R_s (R_s - R_L) / (R_L + R_s)^3 \qquad (2.6)$$

Equating this to zero we find that the maximum transfer of power occurs when $R_L = R_s$ and under these conditions

$$P_L = P_{max}/4.$$

While it may seem somewhat disappointing that the best we can hope for is the transfer of 25% of the maximum available power from the signal source to the load, R_L, it is perhaps somewhat reassuring to realise that this is 50% of the power actually being generated by the signal source under the conditions of the circuit in fig 2.7.

A similar argument may be applied to the circuit of fig 2.5, in which the resistance R_s is replaced by the source impedance and the resistance R_L by the input impedance of the signal converter. In the general case of complex impedances it is found that, for maximum power transfer of an ac signal, the source impedance must equal the complex conjugate of the load impedance. This leads to one of the fundamental rules of electrical signal transmission - that for the maximum transfer of power between a signal source and a load, the impedances of the source and the load must be matched, where "matching" implies equating resistances or the conjugates of impedances, depending on the nature of the signal. This rule is of particular importance in connection with the measurement of transducer output signals because the limit of sensitivity of the transducer is largely determined by the smallest amount of energy which can be measured at the input of the signal converter circuit. Clearly this will represent a higher proportion of the transducer's capability if the impedances are matched than if less than 0.005% of the transducer's output is actually usable at the signal converter's input. Some examples of common transducers are given in table 2.2, along with their characteristic output impedances and typical signal levels.

In spite of this demonstration that the maximum power transfer occurs when the

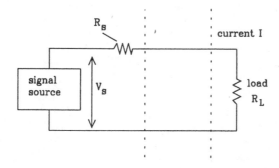

Fig 2.7 *Equivalent circuit for the source of fig 2.6 connected to a load*

impedances at the ends of a transmission line are matched, it must be understood that it is not always necessary or desirable to operate under conditions of maximum power transfer. For example, if we wish to measure a dc voltage generated by a source, then using a measurement device of equal impedance will result in current being drawn through the source series resistance, with the consequent potential drop across that resistance introducing an error into the voltage measurement. In fact under these conditions the voltage at the input terminals of the measurement device will be half of that being generated by the source. To measure 100% of the source voltage we would need a measuring device of infinite input impedance, and in this case no power would be transferred from the source to the measuring device. In general it is desirable to measure dc voltages with a measuring device having a much larger input impedance than the series impedance of the source - this allows the measurement to be made with the highest accuracy and without the necessity of correcting for voltage drops across the sources series impedance. BUT, because the energy transfer involved is very small, there is the possibility of interference by other small energies which may find their way into the measuring device (see section 2.4).

On the other hand dc current measurement would be better accomplished using a measuring device having a very low input impedance compared with the source's series impedance - for under these conditions the current flow is maximised. Again little power is transferred to the measuring device because, although current flows through it, the potential drop which results is small and little power is dissipated in the measuring device - most of the power being dissipated in the source series resistance.

While ac signals at low frequencies can generally be treated in much the same way as dc signals, at high frequencies (>1 MHz) there is not quite as much choice. If impedances at the ends of a transmission line are not matched, then high frequency signals may be reflected from the terminations and their energy dissipated in the conductors of the transmission line.

Table 2.2   Examples of common transducers for laboratory measurement purposes

transducer	output impedance	typical signal range
hot filament	10^1 R	0.1 - 10 V
ion chamber	10^7 R	0.1 - 100 nA
air thermocouple	10^2 R	1 - 100 nV
pellistor	10^2 R	1 - 100 mV
pH electrode	10^9 R	-1 - +1 V
photodiode	10^2 R	1 nA - 1 mA
photomultiplier	10^8 R	1 - 1000 nA
piezo crystal	10^4 R	1 - 100 mV
pyroelectric fet	10^3 R	1 - 1000 mV
strain gauge	10^4 R	0.1 - 10 mV
thermistor	10^4 R	1 - 100 mV

(This can cause high voltages to be developed - higher than those being applied by the source - and these can be dangerous if breakdown of the cabling insulation occurs.) At frequencies above 100 MHz virtually every signal handling conductor needs to be treated as a transmission line and to have its end impedances matched.

2.4 Noise and interference

Two factors which are of importance in determining the limit of sensitivity of a transducer system are the often confused phenomena of noise and interference. Noise consists of fluctuations in the signal level and may be a feature inherent in the nature of the signal or superimposed on the signal by circuit elements in the source and measurement circuits. In properly designed systems the noise introduced into the system by the measurement electronics should be small compared with that originating within the transducer and its associated series resistance, so that noise may normally be regarded as a property of the signal source. Interference on the other hand consists of fluctuations of the signal level caused by the addition of spurious signals from sources not directly involved with the measurement process. One of the commonest forms of interference is the induction of small currents in the signal transmission lines by the interaction of electric and magnetic fields with the conductors which make up the transmission lines.

The measurement system and sources of noise and interference are illustrated in fig 2.8. We will briefly consider each of these problem areas in turn, although it should be appreciated that our purpose is to introduce the nature of the problem rather than to consider the subject in great detail. Firstly there is noise associated with the nature of the signal itself. Actually there are two kinds of noise which should come under this heading, but one of these - fluctuation of the physical property being probed by the transducer - is more conveniently described as background signal variation. This will not be further considered in the present context, as proper (although sometimes very time consuming) design of the physical system involved can reduce this source of noise to insignificant levels. The more limiting source of signal noise is "shot noise", which is the inherent fluctuation of the signal level due to the statistical nature of the electric current (ie. charge movement) produced within the transducer.

The nature of shot noise may be most readily understood by considering a current, I, measured by detecting the number of electrons, x, arriving at the signal converter input in time t. As the rate of arrival is x/t it follows that the current should be xe/t, where e is the electronic charge. However, the statistical variation in the number of electrons arriving in time t is given by the variance in x, which we can take as $x^{1/2}$, so that the number arriving is more properly written as

$$x \pm x^{1/2}$$

and the current as

$$(x \pm x^{1/2}) \, e \, / \, t$$

The noise (ie. the statistical variation) in the measured current may be quantified by its

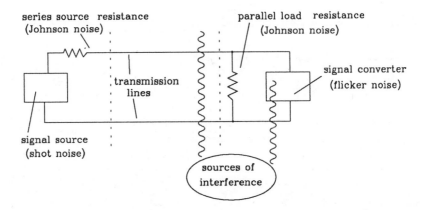

Fig 2.8 *Equivalent circuit for a typical transducer measurement system, illustrating the major contributors to noise and interference signals: the signal source, the source series impedance, the transmission lines, and the signal converter circuit.*

variance, so that as a fraction of the mean current the noise is

$$(x^{1/2} \, e/t)/(xe/t) = x^{1/2} \, /x = x^{-1/2}$$

If the number of electrons arriving in time t is now increased to 4x, the current may now be written

$$(4x \pm 2x^{1/2}) \, e/t$$

and the fractional noise in the current is reduced to

$$2x^{1/2}/4x = x^{-1/2}/2$$

In each case the noise is an example of shot noise, and the calculations illustrate that the shot noise is reduced by increasing the number of physical events contributing to the measurement - in this case electrons recorded by the signal converter.

It is instructive to consider one further example in which the mean current remains xe/t, but the time interval over which the arrival of electrons at the signal converter input is recorded is increased from t to 4t. In this case the number of electrons to arrive may be written as $4x+2x^{1/2}$, and the current becomes

$$(4x \pm 2x^{1/2}) \, e/4t$$

The fractional noise in the current is now

$$2x^{1/2} \, /4x = x^{-1/2} \, /2$$

illustrating the important rule that it is the number of physical events recorded which governs the shot noise, rather than the rate at which the events occur. Often shot noise effects may be reduced by making measurements over a longer period of time.

Shot noise effects are of particular importance in event counting systems, such as single photon spectrometers or single ion mass spectrometers, but other sources of statistical fluctuation in signal level are to be found in more humble systems. For example, the gain of a typical photomultiplier tube may be quoted as $3.0*10^6$, but this represents an average gain for electrons ejected from the photocathode. Many photoelectrons will be multiplied by smaller values, and many others by larger values. Thus the anode current of a photomultiplier exposed to a steady light level will show variations in magnitude due to shot noise, and these variations will represent a larger fraction of the mean signal level, and so appear as a higher noise level, as the light level is reduced.

Shot noise is a characteristic which arises because of the discrete nature of electric current (or photons), and so its effects are discernible for all devices which utilise currents. The magnitude of the shot noise inherent in the flow of a current is given by

$$<I_n> = (2eI\Delta f)^{1/2} \text{ A} \tag{2.7}$$

where   e is the electronic charge
   I is the average dc current, and
   Δf is the frequency interval (bandwidth) over which the shot noise current is required.

With Δf in hertz, eqn (2.7) becomes

$$<I_n> = 1.789*10^{-10}(I\Delta f)^{1/2} \text{ A}$$

The noise level is a function of the frequency interval rather than the frequency, and for a given frequency interval is independent of the frequency. Consequently shot noise is known as white noise, being present at the same intensity at all frequencies. However, because bandwidths are generally a larger number of hertz for circuits handling high signal frequencies than for those handling low frequencies, shot noise produces its greatest practical effect in high frequency systems. For this reason high frequency analog signal handling systems often employ bandwidth narrowing techniques to minimise noise levels.

A second important source of noise is the thermal excitation of electrons in resistive elements of the circuits associated with signal handling. This is called Johnson noise and appears as a fluctuating voltage across a resistive component - whether or not that component is passing a current. Its magnitude may be expressed as the mean value of the noise voltage, $<V_n>$, within a specified frequency range, using

$$<V_n> = (4kTR\Delta f)^{1/2} \text{ V} \tag{2.8}$$

where   R is the resistance of the element

k is the Boltzmann constant

T is the absolute temperature of the element, and

Δf is the frequency interval over which the noise voltage is required.

With R in ohms and Δf in hertz, eqn (2.8) becomes

$$<V_n> = 1.3*10^{-10} (R\Delta f)^{1/2} \text{ V at } 300 \text{ K}$$

As was the case with shot noise, the Johnson noise level is a function of the frequency interval rather than the frequency, so it is a white noise and contributes the same noise level in each equal frequency interval, again producing the greatest practical effect in larger bandwidth, high frequency circuits. It is clear from eqn (2.8) that its effects are minimised by the choice of low value resistances for circuit elements handling very small signals (such as the signal converter input circuit), and by restricting the range of frequencies accepted by the signal converter. Unfortunately it is not always possible to implement these desirable features into a practical converter circuit, and compromise is sometimes required.

Shot noise and Johnson noise are factors which are inherent in the physical nature of current and the resistance to current flow respectively. However, there is a third source of noise which can be distinguished and which arises from the circuit elements used to handle electrical signals (or indeed any other kind of signals). This type of noise is called flicker noise or 1/f noise, and has an inverse frequency dependence - being most troublesome for low frequency or dc handling systems. Flicker noise appears to originate from imperfections in manufactured circuit elements, such as resistors and transistors, and its magnitude increases with the current flowing through the device (unlike shot and Johnson noise).

Flicker noise is normally considered on a purely empirical basis and for modern circuit elements, such as transistors and integrated circuit amplifiers, is lumped together with shot noise arising from the small currents used in the bases or gates of component transistors. This form of noise may be regarded as consisting of a noise voltage and a noise current present at the input of the amplifier, and so subject to the same degree of amplification as the signal being processed (the noise is said to be "referred to input"). Circuits accepting signals from low impedance sources (<1 k) tend to have their output noise level dominated by the voltage noise at the input, while those with high impedance inputs (>1 M) have their output noise level determined principally by the input noise current flowing through the high input impedance.

The noise levels in standard IC amplifiers can be surprisingly high, the values for the famous 741 op-amp (see chapter 3) being about 70 nV $Hz^{-1/2}$ and 1 pA $Hz^{-1/2}$ (at a frequency of 10 Hz). Manufacturers of "low noise" IC amplifiers publish data sheets containing details of the noise levels associated with their devices, and for typical high quality amplifiers figures of the order of 10 nV $Hz^{-1/2}$ and 0.1 pA $Hz^{-1/2}$ (at a frequency above 10 Hz) are to be expected, although these figures may rise at lower frequencies. This corresponds to a total noise level of more than a microvolt voltage noise and more than 100 pA current noise when integrated over the frequency range 0.1 - 10 Hz. It must be remembered that these figures are referred to the input and so apply before amplification. Newer devices may be expected to

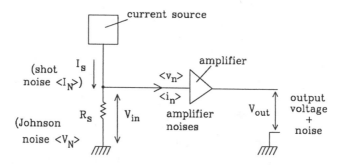

Fig 2.9 *A hypothetical system illustrating four sources of noise: shot noise in the source current; Johnson noise in resistor R_s; and the amplifier voltage and current noises.*

improve on these figures by about an order of magnitude, but nevertheless such noise levels still present a fundamental limitation to the precision with which small, low frequency signals may be measured.

The three important forms of noise outlined above are essentially independent of one another and the total noise level of a system can be estimated from the square root of the sum of the squares of the individual contributions. Generally only one of the noise terms will dominate the total noise, but an example showing all four terms is illustrated in fig 2.9. A current source is passing its current, I_s, into a resistor, R_s, producing a voltage drop which is to be amplified by a voltage amplifier. The noise levels can be considered independently of any gain we ascribe to the amplifier. There are four noise sources:

1) Shot noise, $<I_N>$, in the signal current, I_s.
2) Johnson noise, $<V_N>$, in the resistor R_s.
3) Amplifier voltage noise, $<v_n>$ (flicker dominated at low frequencies).
4) Amplifier current noise, $<i_n>$ (also flicker dominated at low frequencies).

The two current noise terms produce voltage noises by passing through R_s , and the total noise may be written

$$<V_T>^2 = (<I_N>R_s)^2 + <V_N>^2 + <v_n>^2 + (<i_n>R_s)^2$$

While amplifier noise terms can generally be derived from manufacturers' data, some care is needed in estimating the correct values of the shot and Johnson noise terms. For example, the shot noise in a photomultiplier anode current is very much greater than that given by eqn (2.7), because the anode current is a highly amplified version of the photocathode current and contains at least the same fractional noise as that present in the photocathode current. Thus for a tube operating with a gain of 10^6 and an anode current of 10^{-6} A the shot noise in the anode current for a bandwidth of 1 Hz is actually $6*10^{-10}$ A, or 0.06%, rather than the $6*10^{-13}$ A, or 0.6 ppm, suggested by the application of eqn (2.7) to the anode current alone. So the calculation of shot noise for currents arising from transducers with built-in gain must be carried out using the unamplified current value. Furthermore although photomultiplier tubes present a very high resistance pathway to current flow, they do so by virtue of a vacuum between their electrodes rather than a resistive material, so that noise arising from thermal excitation of electrons in resistive materials does not contribute to photomultiplier noise and these devices have no Johnson noise component. (Unfortunately they make up for this by thermal emission from the photocathode giving rise to a dark current).

2.5 Minimising interference

Interference is most frequently caused by small currents induced in the signal transmission lines as a result of the interactions between changing electromagnetic fields and the conductors. If the mean value of this induced current is written $<i_i>$, then when this current flows through, say, the input resistance, R_{in}, of the signal converter circuit it results in a voltage variation given by

$$<V_i> = <i_i>R_{in}$$

It follows that interference is most likely to be serious when R_{in} is large. For example, with $<i_i> = 10^{-8}$ A, an input resistance of 1M gives rise to $<V_i> = 10$ mV, while an input resistance of 50R gives rise to $<V_i> = 0.0005$ mV. The importance of the values in relation to the measurement of a 10 mV signal is self-evident.

When the source and the signal converter have low impedances interference is rarely a problem, the precision with which a signal may be measured being limited by noise in the signal and flicker noise introduced by the circuit elements of the converter. Unfortunately many widely used transducers have a very high output impedance and so require signal converters with a high input impedance. To minimise the effects of interference in such systems the first priority should be the elimination of any local sources of interference, as this often significantly relieves the problem. Thus any troublesome local thermostats and motors should be fitted with interference suppressors. However, there are other steps which can be taken to minimise the effects of interference generated outside local control, and most of these steps concentrate on the points at which the interference currents are introduced to the transmission lines and the signal converter input.

The first approach is to minimise the magnitudes of currents induced on the signal transmission lines by keeping these as short as possible, preferably less than a few centimetres long. Secondly, the transmission lines should be shielded (or screened) wherever possible to prevent electromagnetic fields reaching the conductors. While conventional coaxial cable, consisting of an inner conductor and an outer shielding braid connected to ground, is better than nothing, a preferred technique is to shield both conductors of the transmission line - even if one of these is grounded. (Interference currents are generally unaware of the purpose served by the particular piece of conducting material they inhabit.) Shielding should be connected as illustrated in fig 2.10 and grounded at one end only, to prevent induced currents flowing along the shielding and inducing currents in the adjacent transmission lines.

In some cases long lengths of transmission lines are unavoidable and shielded conductors

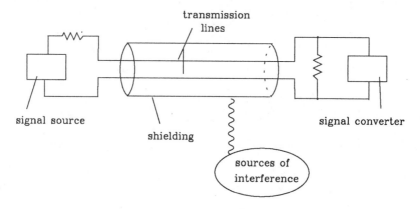

Fig 2.10 *The use of shielding to minimise the induction of interference currents in transmission lines.*

(which are never perfectly shielded) are inadequate. In such cases the easiest solution is often to move a part of the signal converter input circuit from the converter end of the line to the signal source end. This can be achieved by including an impedance converter circuit (see chapter 3) as close as possible to the signal source, and converting the high impedance output of the source to a low impedance output of the impedance converter, as illustrated in fig 2.11. If the signal converter input is now a low impedance input, interference pickup on the transmission line is likely to be small even on many metres of transmission line.

All that is needed to achieve successful interference pickup is a conducting pathway from the laboratory environment to the input circuitry of the signal converter. Most sensitive signal converter systems are likely to be housed in metal cases, although conducting paints for plastic cases and cases constructed of plastics laminated with a conducting layer are also available. In sensibly designed systems, these cases can provide their own shielding for the

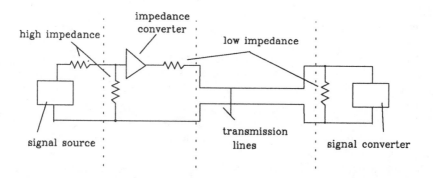

Fig 2.11 *The use of an impedance converter positioned close to a high impedance signal source to provide a low impedance output for connection to long transmission lines.*

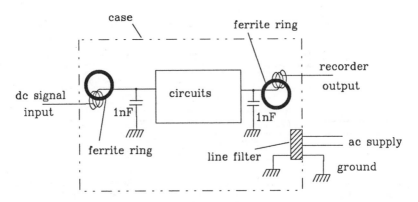

Fig 2.12 *Typical techniques for minimising the introduction of interference signals to instrumental circuits by suppressing leads entering the instrument case. (Both recorder lines should receive the same treatment, as should both input lines - even where one is nominally a ground line.)*

wiring inside, but it should not be forgotten that in many systems conductors other than the transmission lines pass through the case walls. In particular a mains (or line) supply may be connected to miles of unshielded conductor acting as a magnificent antenna for all the interfering signals in town. Wrapping a 1 metre length of plastic covered wire around the mains lead of an oscilloscope and viewing the signal picked up on the wire will demonstrate the magnitude of the mains-borne interference problem in any laboratory. Usually there is no shortage of interference in the 1-100 MHz region. A wide selection of rf line filters is available at relatively low cost and many of these are reasonably effective at preventing interfering signals being radiated from mains conductors within the instrument case. Unfortunately ground wires can still be a serious source of interference, and it may be helpful to ensure that only one connection is made between the ground points of individual circuits and the external ground provided by the mains ground wire.

Low level dc supply lines and chart recorder output wires are other, sometimes neglected, sources of interference pickup, and the solution is often to be found in screening the cables and installing ferrite hoop filters on the wires inside the case and as close to the entry point as possible, followed immediately by a small value capacitor, say 1 nF, as illustrated in fig 2.12. (Incidentally, using a ferrite hoop without a following capacitor can make matters worse rather than better, as the hoop may become an effective radiator of interference.) It should also be borne in mind that the multiphase wiring of modern buildings, fluorescent lighting and the high currents used in much modern laboratory equipment result in a number of perhaps unexpected phenomena. For example, fluorescent lighting produces a significant amount of radiation at double the mains frequency. Furthermore, it is not uncommon to find dc level differences of several hundred mV between the ground pins of two mains outlets in the same laboratory, and to find different patterns of rf coming from each outlet. It is usually easier to tackle mains and ground-borne interference if the mains connections of each unit of a system can be made to a single outlet.

2.6 Signal-to-noise ratio

A parameter which is most useful in defining the quality of a complete measuring system is the signal-to-noise ratio of a specified signal. For example, the limit of measurement for many systems is defined as the signal level which can be measured with a signal-to- noise ratio of 2:1 (or just 2). The signal-to-noise ratio of a system is normally determined at the output of the signal converter; however, it is convenient to introduce the quantity formally at this point now that we have given some thought to the origins of noise and interference. In the present context the word noise is being used to mean the fluctuations in the output of the measuring system, deriving from all sources, and is quantified as the variance of the output signal. The signal-to-noise ratio (SNR) may be defined as the ratio of the mean output of the measuring system to the variance of that output, ie

$$SNR = <V_S>/<V_N>$$

where $<V_S>$ and $<V_N>$ are the mean values of the signal voltage and the noise voltage respectively. An analogous expression may be written for the SNR of current output systems, although these are less common. The output in question may be a nominally constant signal level (such as the value recorded by a pH meter with the probe in a buffer solution), or a difference between values of a signal level which changes while the measurement is being made (the height of a chromatographic peak for example). The contributions of signal, noise and interference to the output of a dc measuring system are illustrated schematically in fig 2.13.

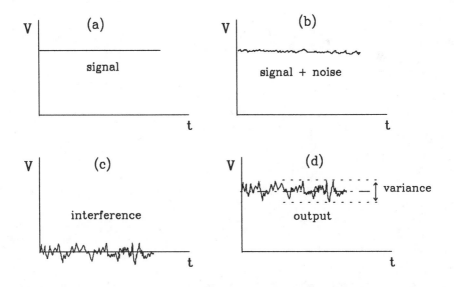

Fig 2.13 *The effect of adding to a constant level signal, (a), contributions of noise, (b) and interference, (c), to produce the kind of signal, (d), usually encountered in laboratory measurements.*

Signal-to-noise ratios are often expressed using the dB scale, in which case

$$SNR = 10 \log_{10}(<V_S>^2/<V_N>^2) \ \ dB$$

but whichever system of units is used it is important to specify the bandwidth over which the SNR applies. For the measurement of a signal of closely defined frequency, the use of a measuring system with a wide frequency response allows an increase in the observed value of $<V_N>$ without any improvement in $<V_S>$, thus producing a degraded SNR. The situation is particularly critical for the measurement of dc signals because of the contribution of 1/f noise. While it may be tempting to assume that f has no relevance to dc measurements, think again. One generally requires laboratory measurements to be completed within a finite time, and even if we allow 100 s for a particular measurement, this corresponds to an f of 0.01 Hz - so a bandwidth of 1 Hz for the measuring device is 100 times greater than is needed and allows $<V_N>$ to be greater than necessary.

The optimistic reader may feel that these problems of noise and interference have been somewhat laboured. However, there is a good reason for this. In later chapters we will be considering the recording of signal levels by computer. A computer can collect the value of a signal at a particular microsecond, while a pen recorder records a time-averaged signal (typical pen recorder response time is of the order of 0.5 s) so that any rapid fluctuations in a signal level are usually not apparent on a chart record. A signal which records on a chart recorder as a noiseless line may, when read twice per second by computer, appear as a most disappointing collection of apparently random numbers.

2.7 Control signals

Most of the signals we have encountered from measuring devices are also used for control functions or for the production of physical effects required in a laboratory. Examples of some common uses of signals for these purposes are given in table 2.3. The principal differences between the characteristics of signals associated with measurement and control are the magnitudes of the signals and the impedances of the devices which generate or utilise them. Thus the photomultiplier detector of a spectrometer may produce an output of a few microamperes when a strong signal is being recorded, whereas the light source of the same spectrometer may require a current of several amperes, supplied with a potential of many tens of volts.

Signals at the higher levels common for control and power functions of instrumentation are not normally subject to interference, although often responsible for causing interference in other systems. However, noise (ie. fluctuations of the signal level) can present serious problems, particularly if the fluctuations are likely to affect some sensitive measurement transducer. Dc signals at low voltages (<100 V) are relatively easy to stabilise and filter using voltage regulator ICs and simple RC filters, but at the high voltages generally required as the bias supplies for photomultiplier and channeltron detectors, the removal of noise is much more difficult. Most commercial power supplies in this category are rated for their noise and ripple (ie. residual variation at the ac frequency used to generate the high voltage) levels, and these levels are likely to increase as more current is drawn from the supply. Looking at the

Table 2.3   Examples of controlled devices

device	operating signal	output
discharge lamp	100 V, 5 A, dc	uv light
filament	10 V, 1 A, dc	red light
heater	mains 5 A	heat
LED[a]	5 V, 5 mA, dc	low-level light
loudspeaker	10 V, 1 A, ac	sound
microwave source	10 V, 100 mA, dc	X-band microwaves[b]
motor	24 V, 2 A, dc	motion
pump	mains 1 A	fluid pressure
stepping motor	5 V, 1 A, pulses	speed/position
solenoid valve	12 V, 100 mA, dc	fluid flow
ultrasonic source	12 V, 25 mA, ac	ultrasound
X-ray generator	8 kV, 100 mA, dc	X-rays

[a] light emitting diode. [b] low power modules.

noise level on high voltage lines can be difficult as most oscilloscopes have maximum input voltages of 500 V or less (whereas photomultipliers generally require 1000 - 2000 V bias), and it is usually necessary to divide the examined voltage down using a high voltage dividing probe.

Ac power signals derived from the mains supply are most likely to be used for the operation of heaters, lights and simple motors, whereas ac signals at higher frequencies are used for more exacting requirements, such as precision conductivity measurements, magnetic resonance systems, quadrupole mass spectrometric analysis and rf heating. Non-sinusoidal oscillating signals are also popular; for example, triangular waveforms are useful in providing short duration, constant acceleration motions, and square waves are the ideal repetitive on/off signals for lamps and mechanical devices. On/off signals with a variable on/off ratio are useful for driving dc motors at varying speeds without the loss of torque which accompanies voltage reduction, and motor controllers are available for this purpose.

Ac signals up to mains voltage can be turned on and off using solid state relays or triacs, and opto-isolated versions of both devices are available to preserved high reliability insulation between the switching circuit and the switched signal. Triacs can also be used for controlling power in ac circuits, although generally only if the load is resistive (eg. a lamp or heater). Dc signals may be turned on and off using power transistors or regulator ICs, and the development of VMOS and DMOS power fets enables this technique to be used for signal levels beyond 1000 V. Controlling dc levels is most readily accomplished using variable level regulator ICs, although for voltages greater than about 100 V it is necessary to use a power fet and to control its gate voltage. Most of the necessary devices can be found in the catalogues of electronic component suppliers, and many suppliers will also be able to supply the relevant data sheets on request.

CHAPTER 3

THE ELEMENTS OF ANALOG SIGNAL HANDLING

Most transducers and laboratory signal sources yield output signals the magnitude of which carries at least a part of the information required in the measurement process. Such signals are called analog signals. It follows that most signal converter systems are required to handle analog signals, at least in the input stages of the converter circuit. One of the most widely used circuit elements for handling analog signals is the amplifier, which, as its name implies, is intended to convert a small signal into a larger one, often to overcome the inconveniently small magnitude of the signal generated by a transducer. Early amplifiers were constructed using thermionic valves, which were physically large and consumed large amounts of power - most of which was wasted as heat. Later amplifiers were constructed using transistors, and many special purpose amplifiers (such as VHF, Very High Frequency, amplifiers) still need to be assembled from a number of these "discrete" devices. However, in the majority of cases amplifiers can be more easily constructed using integrated circuit elements called operational amplifiers (op-amps for short). Op-amps have in fact become the basic ingredient in nearly all aspects of analog signal handling and we shall concentrate our discussion on these devices.

3.1 Op-amps

Operational amplifiers are supplied in a variety of packages, including DIL packages, TO-8 and TO-99 cans. The symbol used for the operational amplifier in circuit diagrams is shown in fig 3.1, along with the names of the most important connections to the package. The devices are normally powered by dual supplies of ± 5 to ± 15 V, although some single supply op-amps are also available. Op-amps have two inputs (an inverting (-) input and a non-inverting (+) input) and one output, and "operate" on the difference in potential between the two inputs - producing an output proportional to that potential difference. An ideal operational amplifier has the following characteristics:

a) it produces an output voltage which is directly proportional to the voltage difference between the non-inverting and inverting inputs.
b) the constant of proportionality in a), ie. the voltage gain, is very large - essentially infinite.
c) the bandwidth of the amplifier is infinite.
d) the input impedance at both inputs is infinite.
e) the output impedance is zero.

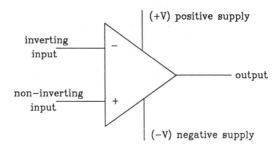

Fig 3.1   *The standard circuit symbol for an operational amplifier, showing the functions of the essential connections. In subsequent figures the power supply connections will be omitted.*

Not surprisingly, available op-amps do not actually achieve these ideals. Table 3.1 shows the actual properties of some popular op-amps, and allows us to introduce some of the imperfections of practical devices which need to be taken into account in the design of instrumental systems.

The voltage gain of a practical op-amp is never infinite. Typical values range from 10^4 to 10^7. These values actually refer to the open-loop voltage gain, ie the voltage gain of the op-amp when the output signal is not fed back (or looped) to one of the inputs. In most applications some part of the output signal will be fed back to the inverting input, and as we shall see later, this results in the effective voltage gain of the circuit being quite low (typically 1 - 1000) but stable and controllable. For this reason the fact that the open loop gain is not as large as infinity does not impose a serious limitation on the use of op-amps, although the frequency dependence of the open loop gain is quite a different matter (see below). Of course, under no circumstances can the value of the output voltage exceed the supply voltage available to the op-amp; once either the positive or negative supply voltage is reached the output is said to be saturated. (Some op-amps cannot provide an output as large as the supply voltage.)

The input impedance of practical op-amps is also somewhat lower than infinity, so that some current is actually drawn from a signal source connected to an op-amp input. As impedance is a frequency dependent property, manufacturers normally specify the dc input resistance, although for frequencies below 1 MHz this is likely to be adequate for most calculations. For available bipolar op-amps (op-amps constructed using bipolar transistors) the input resistance is typically greater than 1M, although this is not high enough to match the ouput impedance of many important transducers (eg. pH electrodes). Fortunately a type of op-amp which uses field-effect transistors for its input stages (known as a BIFET op-amp) is now readily available, and op-amps in this category can have input resistances of 10^{12} R or more.

The ouput impedance of common op-amps is low enough for most purposes, typically being <10 R and sometimes <1 R for dc signals. However, the bandwidth of available

Table 3.1 Principal characteristics of some popular operational amplifiers

Device no.	741	531	OP07	071	LH0032
Type	- - bipolar - -			BIFET	FET
Input resistance /R	2 M	20 M	33 M	10^{12}	10^{12}
Open loop voltage gain /10^6	0.2	0.06	3.9	0.2	0.003
Input offset voltage /mV	2	2	0.06	3	2
temp.co. /μV/°C	5	2	0.5	25	
Input bias current /nA	80	400	2.2	30 pA	5 pA
Input offset current /nA	20	50	0.8	3 pA	-
temp.coef. /nA/°C	0.5	0.6	0.012	-	-
small signal bandwidth /MHz	1	1	-	3	70
slew rate /V/μs	0.5	35	0.17	13	500
full power bandwidth /kHz	10	500	100	150	7 MHz
CMRR /dB	90	100	120	76	60
Maximum output current /mA	20	10	15	10	15

operational amplifiers does present something of a problem - not least because it is often difficult to determine precisely over what frequency range an op-amp is likely to function. The relationship between bandwidth and the application of the op-amp compels us to defer discussion until we have examined some op-amp circuits. For the present it will be sufficient to note that while early op-amps were essentially low frequency devices (operating from dc up to about 100 kHz), some of the op-amps now available will function up to 100 MHz. Sadly those op-amps which can operate above a few MHz tend to have a number of disadvantages, such as a high power consumption (causing temperature rises), high noise levels and a desire to oscillate. Consequently the use of these devices is desirable only when high frequency operation is essential, and in such cases a high frequency op-amp such as the NE5539 may be used. At lower frequencies the more mundane (and much cheaper) op-amps with modest bandwidths provide far less trouble than their high speed relations.

Unfortunately there are a number of other imperfections associated with practical op-amps, and some of these are sufficiently important to be considered before we look at the uses of these devices. Two of the more serious imperfections are called the input offset voltage and input offset current. These terms refer to differences between the inverting and non-inverting inputs of the op-amp. For example, when both input connections are connected together and grounded (as shown in fig 3.2a), a voltage appears at the op-amp output. The value of this output voltage divided by the voltage gain of the circuit is called the input offset voltage. Its value usually sounds quite small for normal (bipolar) op-amps, typically <0.1 mV for newer designs such as the OP07 family, although it can be significantly larger (up to 10 mV) for FET input op-amps. However, input offset voltage is subject to the same voltage gain as the difference signal applied between the two inputs, so that when very small signals are being amplified the input offset voltage can be an inconvenience. Some operational amplifiers are provided with two connections which can be used to balance or "offset null"

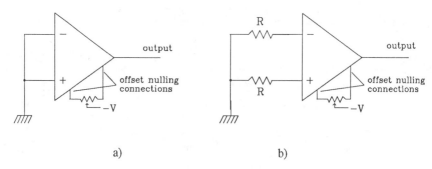

a)   b)

Fig 3.2 *An op-amp connected for measurement of a) input offset voltage, and
b) input offset current. Note that some op-amps require the offset
voltage nulling potentiometer to be connected to the positive supply
rather than as shown.*

the input circuits with a single preset potentiometer. An alternative arrangement which works
for any op-amp is considered in section 3.5.3.

To make matters even worse, the input offset voltage varies with temperature, the
variation being known as the input offset temperature coefficient or drift, and being specified
in mV/°C. Newer bipolar op-amps may have relatively low coefficients, but FET input op-
amps have rather large ones, typically 25 µV/°C. This can be the source of some
inconvenience unless some thought is given to keeping op-amps handling very small signal
voltages away from sources of heat - such as the heat sinks of power supplies. Furthermore,
when an op-amp is driving a low impedance load, the output current may be large enough to
cause heating of the amplifier package, which in turn can exacerbate drift. To minimise
problems from this source it is preferable to operate high gain amplifiers with output loads of
10 k or more.

The second input offset problem is the input offset current. If the input offset voltage is
nulled so that when both inputs are grounded the output voltage is zero, and the op-amp is
then connected as shown in fig 3.2b, it is found that the output voltage is no longer zero. This
arises because the two op-amp inputs draw slightly different currents, called input bias
currents, through the resistors, R, and so the inputs are at slightly different potentials - thus
giving rise to an output. There is little that can be done to overcome the effects of the input
offset current, beyond selecting an op-amp which has a small value for this parameter.
However, the effect does remind us that the op-amp inputs do draw bias currents, and that
when these bias currents pass through resistances before reaching the inputs, the voltage
drops across those resistances may contribute to a voltage difference between the op-amp
inputs - thus giving rise to an output. For this reason it may be important to ensure that the
total resistances of circuits connected to the two inputs of op-amps are approximately equal.

3.2 Feedback systems

As indicated above operational amplifiers are not intended to operate with an actual
voltage gain as large as the open loop gain. Op-amps are provided with such high gains to

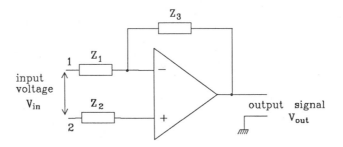

Fig 3.3 *Generalised op-amp feedback loop circuit. The op-amp adjusts its output in an attempt to make the voltage difference between the (+) and (-) inputs zero.*

provide an extremely versatile building block with which to construct a wide variety of circuits. In practice operational amplifiers are used in circuits which feedback some of the op-amp's output to one of its inputs. The starting point for our brief survey of op-amp applications is the feedback loop circuit of fig 3.3. In this circuit an input voltage signal is applied between points 1 and 2. The input bias currents flowing through the impedances Z_1 and Z_2 to the two inputs of the op-amp are usually small, so that the voltage drop across the impedances can be neglected (unless the impedances are very large, >1 M, or the signal very small, <1 mV). Thus the potentials at the inverting and non-inverting inputs may be taken as equal to those at 1 and 2 respectively. The output voltage of the op-amp (with respect to ground), V_{out}, is proportional to $V_{in}=V_2-V_1$, the voltage difference between the inverting and non-inverting inputs. In the absence of the feedback impedance, Z_3, the output voltage would try to reach a value given by the product of the differential input voltage and the open loop voltage gain, G_{ol}, ie.

$$V_{out} = G_{ol}*V_{in}$$

(Of course, it would not succeed if this product exceeded the supply voltage.) However, the presence of Z_3 allows current to flow between the output and the inverting input, and this current opposes that flowing through Z_1 to the inverting input. Because the voltage gain of the op-amp is so large, any finite voltage difference between the inverting and non-inverting inputs would give rise to a saturated output voltage, which in turn would cause current to flow through Z_3 in such a direction as to remove the differential input voltage. The practical result of this is that the output voltage settles to the level required to make the voltage difference between the inputs equal to zero. In the case of fig 3.3 this occurs when

$$V_{out} = V_{in}*(Z_3/Z_1) \qquad\qquad\qquad (3.1)$$

While Z_2 does not enter into the equation for the output voltage, we would choose the real part of Z_2 to be equal to the dc resistance seen by the op-amp from the inverting input (ie. the real part of Z_1 in parallel with Z_3) so that the small potential drops associated with the input bias currents were equal at both inputs.

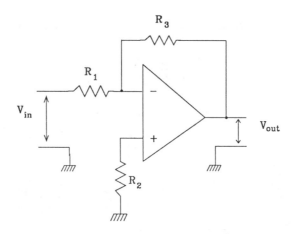

Fig 3.4 *The basic inverting amplifier configuration.*

3.3 Basic amplifier configurations

The general principles of the feedback loop considered above are relevant to loops containing any type of feedback impedance. Most straightforward amplification applications involving dc or low frequency ac signals require the output signal to be directly proportional to the applied input signal, and in this case the feedback elements are purely resistive. There are three basic types of amplifier circuit which utilise resistive feedback, and these are illustrated in figs 3.4 - 3.6.

Figure 3.4 shows a schematic diagram for the basic inverting amplifier configuration. The circuit acts as a voltage amplifier with a voltage gain (from eqn 3.1) of $-R_3/R_1$ (ie. the sign of the output voltage is different from that of the input signal). R_2 should be chosen so that its value is approximately $R_1R_3/(R_1+R_3)$ to minimise errors due to input bias currents. The amplifier is suitable for amplifying signals ranging from a few mV (positive or negative) to a few volts, subject to the condition that the amplified output does not exceed the supply voltage, and is normally used to provide voltage gains of between 1 and 1000, although higher voltage gains can be achieved. While the resistance values are not critical, it is common practice to use a value of around 10k for R_1 with bipolar op-amps, and 100k - 1M with FETs and BIFETs. Higher values can be used, particularly with FET input op-amps, although there is no escape from the basic limitation of this configuration - that it has an input impedance characteristic of the resistance network rather than of the op-amp itself. This arises because R_2 is grounded on one side, effectively maintaining the non-inverting input at ground potential, so that the inverting input potential is always maintained very close to ground (ie. within a few microvolts) by the action of the feedback loop. In fact the inverting input of the inverting amplifier configuration is often described as being a "virtual ground". The applied input signal therefore sees R_1 as connected to ground potential, and the input resistance of the circuit is R_1. This is generally a much lower value than the characteristic input impedance of the op-amp, and the inverting configuration amplifier is therefore inappropriate for the amplification of signals from high impedance sources.

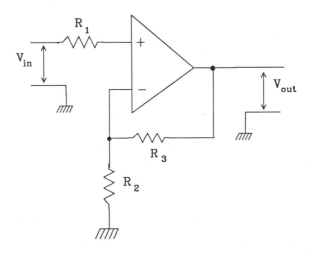

Fig 3.5 *The basic non-inverting amplifier configuration.*

An alternative configuration which overcomes the problem of a restricted input impedance is the non-inverting amplifier shown schematically in fig 3.5. The principal difference between this configuration and the previous one is that neither input is now maintained at ground potential. The non-inverting input will be at the applied signal potential, and the inverting input will, by virtue of the feedback loop, be maintained at virtually the same potential. Thus the impedance seen by the signal source is R_1 (which is normally insignificant) in series with the input impedance of the op-amp (which is typically $>10^7$ R). The signal fed back to the inverting input now derives from the voltage divider formed by R_2 and R_3, and as this must equal V_{in}

$$V_{in} = V_{out} * R_2 / (R_2 + R_3)$$

Rearranging, we find that the ouput voltage is given by

$$V_{out} = V_{in} * (R_2 + R_3) / R_2 \qquad (3.2)$$

and has the same sign as the input signal. Again R_1 is chosen to be approximately equal to $R_2 R_3 / (R_2 + R_3)$ to minimise errors due to bias currents.

While the non-inverting configuration may appear to be an ideal amplifier circuit, one of the imperfections of practical op-amps now attempts to spoil an otherwise promising circuit. The same floating of the inputs which allowed the input impedance to be so high, results in both inputs essentially following the applied signal. This is fine for small applied signals of a few mV, but when the applied signal reaches several volts the output voltage tends to become subject to errors known as common mode errors, and the gain of the amplifier changes. A common mode signal is a signal applied to both inverting and non-inverting inputs (in this case the signal at the inverting input is derived from the voltage feedback of the feedback loop). For an ideal op-amp a common mode signal should produce an output of 0 V. The

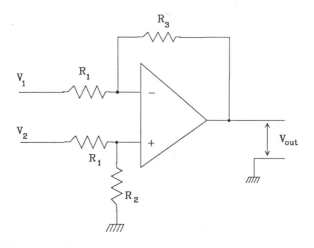

Fig 3.6   *The differential amplifier configuration. The differential input*
          voltage is (V_2-V_1).

ability of a practical op-amp to cope with common mode signals is quantified by the common mode rejection ratio (CMRR), which is the ratio of the output produced by a normal differential voltage signal to the output produced by a common mode signal of the same magnitude. To make matters worse the CMRR is usually quoted in dB and is typically in the range 60-100, which translates to 10^3-10^5 when everything is in volts. Thus in an unfavourable case a common mode signal of 10 V may give rise to the same output voltage as a normal input difference signal of 10 mV. Possibly not a serious problem for most dc applications, but a potential source of distortion for ac signal handling.

Figure 3.6 shows the basic differential amplifier configuration, an arrangement which is useful because the input signal does not need to be related to ground. The output voltage in this configuration is given by

$$V_{out} = (V_2-V_1)*R_3/R_1 \qquad\qquad (3.3)$$

and R_2 is normally chosen to equal R_3 to minimise bias current errors. This configuration is generally adopted when it is desired to amplify a small difference signal (V_2-V_1), which may accompany a significant common mode signal, $(V_1+V_2)/2$, although for most op-amps neither V_1 nor V_2 may exceed the op-amp's supply voltage. The input impedance of this configuration is high, again because the inputs are not virtual grounds, but the price paid for this is the risk of common mode errors. Furthermore, the common mode signal at the op-amp inputs is maintained by the voltages V_1 and V_2 applied through the two input resistors R_1, as well as by the contribution of the feedback loop. For this reason common mode errors are accentuated by errors in the values of the resistors, and to minimise this source of error closely matched resistors should be used for the R_1s and for R_2 and R_3. Resistors with tolerances of better than 1% are normally required, and 0.1% tolerance would be even better.

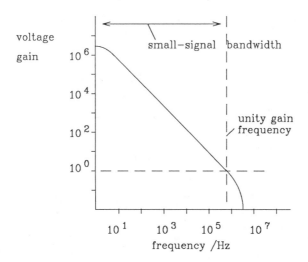

Fig 3.7   *Typical example of the variation of the open loop gain of a 741 op-amp*
with signal frequency. Such curves may be used in designing closed loop
systems for handling small signals. (Conditions: power supply ± 15 V,
output load 2 k resistive, temperature 25°C).

3.4 Bandwidth and slew rate

The three basic amplifier configurations of figs 3.4 - 3.6 may be adopted with most available op-amps and, within the limitations noted, will perform adequately when amplifiying dc voltage signals. However, when the input signal possesses a time-dependent component the ability of the op-amp's output voltage to keep up with changes in the input signal becomes a major consideration. The relevant details on the specification of a particular op-amp are contained in two items of data provided by the manufacturer. The first is a plot of the open loop gain as a function of signal frequency for small signals. (It is not always obvious what manufacturers mean by small signals. In practice it usually means sine waves with amplitudes of a few mV.) A typical gain vs. frequency plot for a general purpose op-amp is shown in fig 3.7. Clearly while the open loop gain is large for dc and low frequency signals, the gain falls dramatically as the signal frequency rises. The important point is that the closed loop gain of an amplifier circuit at any frequency cannot exceed the open loop gain of the op-amp at that frequency. For example, the op-amp specified in fig 3.7 may be used in a circuit operating with a closed loop gain of 100 to amplify small signals with frequencies up to about 10 kHz; but attempts to operate the circuit at higher frequencies will result in a lower gain than that calculated from the input and feedback impedances.

The frequency at which the open loop voltage gain falls to unity (or 0 dB) is properly referred to as the cut-off frequency, although it is often more optimistically termed the small signal bandwidth. As most of the gain vs. frequency plots tend to be straight lines (when the axes are both logarithmic), the term gain-bandwidth product is also commonly used to indicate the frequency capabilities of an op-amp. In our example (fig 3.7) the gain-bandwidth

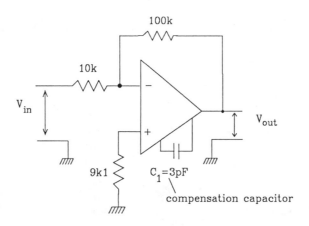

Fig 3.8 *Schematic diagram for a *10 inverting amplifier using an uncompensated 748
op-amp. The compensation capacitance C_1 is chosen according to the
circuit gain and the maximum anticipated signal frequency - 1 MHz in this case.*

product is approximately 1 MHz, so that the op-amp may be used at gains of up to 100 for
small signals up to 10 kHz, or gains of up to 10 for small signals up to 100 kHz, or unity
gain for small signals up to 1 MHz.

In practice the gain vs. frequency characteristic for an op-amp depends on the values of
one or two components known as frequency compensation components. Many op-amps are
available in versions which are "internally compensated," the frequency compensation
components being inside the op-amp package (eg. the 741 is basically a compensated version
of the 748). In these cases the gain vs. frequency characteristic of the op-amp is fixed.
However, uncompensated op-amps require the user to connect his own frequency
compensation components to pins on the op-amp package, and in this case the gain vs.
frequency characteristic is dependent on the value of these components. The data sheets for
such op-amps usually indicate the manufacturer's recommendation for compensation
components for a variety of circumstances, so this additional chore is not particularly difficult.
An example of a *10 inverting configuration amplifier circuit complete with a single
frequency compensation capacitor is given in fig 3.8.

The second important item of data relevant to the signal frequency limitations of an op-
amp is called the "slew rate". This is the maximum rate at which the output voltage can
change, and is quoted in volts per microsecond. Values vary widely, as indicated by the
examples given in table 3.1, and may depend on the supply voltage used. If an op-amp is
amplifying a sinusoidal signal which falls within its small signal bandwidth, and the gain
produced by the feedback loop is increased, then there will come a time at which the
maximum rate of change of the output voltage equals the slew rate for the device. Beyond
this point the amplitude of the output waveform cannot be increased, either by increasing the
gain or by increasing the input amplitude. Attempts to produce a slightly larger output may
appear successful when viewed on an oscilloscope, particularly if the values of the

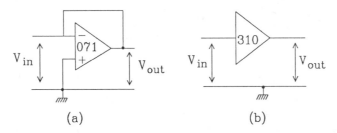

(a)   (b)

Fig 3.9   *a) Schematic diagram for a unity gain voltage follower for which $V_{out} = V_{in}$.*
b) A buffer amplifier with a fixed gain (generally unity); such amplifiers do
not have inverting input connections and so require no external feedback.

resistors setting the gain are small (eg. <1 k), but in reality the output waveform will be distorted and the output amplitude will not be a linear function of the input. The limitation imposed by the slew rate is expressed in its most severe form by the full power bandwidth of the device, which is the maximum frequency at which the output can oscillate with the maximum possible amplitude (usually close to the supply voltages). The full power bandwidth of an op-amp may be two orders of magnitude smaller than its small signal bandwidth, and in general it is only the relatively expensive hybrid devices which can handle large amplitude signals at frequencies above a few MHz. However, as manufacturing techniques have improved, a number of relatively low cost, high speed op-amps have become available, and a typical example is the NE5539 - which has a slew rate of 600 V/ms and a full power bandwidth of 48 MHz. Such speed improvements are usually obtained by making some sacrifice in other areas, generally in the power consumption and input resistance (which is 100 k in the case of the NE5539).

3.5 Practical dc signal circuits

Now that we have surveyed some of the fundamental characteristics of operational amplifiers we can turn to a brief examination of some additional types of op-amp circuits which are useful in the design of instrumental electronics for the laboratory. The circuits described below have all been used for specific applications in the author's laboratory, and so may benefit from some modification before application in different circumstances. However, it is not the intention here to provide detailed information on the do-it-yourself construction of specific circuits, but rather to assist the reader in understanding the basics of what is involved in various signal handling aspects of modern laboratory instrumentation.

3.5.1 Unity gain buffer amplifiers

A fairly common requirement in handling signals in a laboratory is to pass a voltage signal from a high output impedance source (eg. a transducer) into a circuit of a much lower input impedance (eg. a signal converter circuit). This may be achieved using a buffer amplifier, which does not change the magnitude of the voltage signal but affects only the series impedance presented to a subsequent circuit (see section 2.5). One of the most useful types of buffer amplifiers is the voltage follower circuit shown in fig 3.9a. The non-inverting input accepts the input signal, presenting an input resistance equal to the characteristic input

Table 3.2 Typical voltage follower and buffer amplifiers

| Device no. | LM302 | LM310 | LH0033 | LH0063 |
Type	- bipolar -		- - FET - -	
input resistance /R	10^{12}	10^{12}	10^{11}	10^{11}
small signal bandwidth /MHz	10	20	100	200
slew rate /V/μs	10	30	1500	6000
full power bandwidth /kHz	60	250	-	-
maximum output current /mA	20	24	100	250

resistance of the op-amp, typically 10^6 -10^{12} R. The output connection provides both the output signal, at the low characteristic output impedance of the op- amp, typically <10 R, and 100% voltage feedback to the inverting input of the op-amp. The ouput voltage follows the input voltage, although offset nulling may be necessary if high precision is required. Essentially the same effect is produced if resistances are placed in both the (+) and (-) input lines, and this is often advisable to protect the inputs. As the circuit is essentially a special case of the non-inverting amplifier configuration, the voltage follower is subject to common mode errors. On the other hand the op-amp is operated at unity gain, so, if only small signals are involved, the small signal bandwidth of the circuit may be fully utilised.

The use of an internally compensated op-amp allows the voltage follower circuit to be implemented easily for dc and low frequency signals. However, the design of compensation networks for uncompensated op-amps operating at unity gain and high frequencies is particularly tricky. Fortunately a number of IC manufacturers produce dedicated buffer amplifiers in which the feedback and compensation components are contained in the package, and which are therefore very simple to use. Such buffer amplifiers may have no external

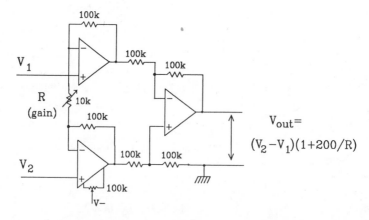

Fig 3.10 *Basic circuit of an instrumentation amplifier with overall voltage gain determined by R, the value set by the 10k variable resistor.*

Table 3.3 Principal characteristics of some instrumentation amplifiers

Device no.	LF352	LH0036	LH0038	ICL7605
Type	FET	bipolar	bipolar	FET
Input resistance /R	$>10^{12}$	300 M	5 M	-
Gain range V/V	1-1000	1-1000	-	-
Input offset voltage /mV	15	1	0.1	0.005
Input temp. coef. /μV/ °C	10	10	0.25	0.1
Input offset current	20 pA	10 nA	5 nA	-
Small signal bandwidth /kHz[a]	7	0.35	1.6	0.01
Slew rate V/μs	1	0.3	0.3	0.5
Full power bandwidth /kHz[b]	25	5	-	-
Noise voltage 0.1-10 Hz /μV	1.3	1	0.2	5
Noise current 0.1-10 Hz /pA	0.01	-	10	-
CMRR /dB[c]	105	100	114	100

Notes: a) varies with gain; stated value is frequency at which nominal
            gain 1000 drops to 500 (ie. -3dB).
          b) unity gain value.
          c) varies with gain, source impedance and frequency! quoted
            value is for gain 1000 and dc.

inverting input connection, so their circuit symbol appears as a modified op-amp symbol as illustrated in fig 3.9b. Examples of some useful buffer amplifiers for a variety of bandwidths are given in table 3.2.

3.5.2 The instrumentation amplifier

An instrumentation amplifier is essentially a circuit with most of the characteristics of an operational amplifier, but with significantly better specifications in terms of gain, noise, stability, input impedance, CMRR and input offset temperature coefficient. Its principal use is in the amplification of small differential signals coming from high output impedance transducers, particularly in situations where a large common mode signal is present. The basic circuit of an instrumentation amplifier is shown in fig 3.10, where it can be seen that non-inverting amplifier elements are used in each input line to maintain the high input resistance of each input. Offset nulling is provided on one of these stages, as this is all that is necessary to trim the output of the whole circuit.

Several manufacturers produce instrumentation amplifiers in a single package, and for most purposes it is better to buy one of these than to attempt to construct one's own from op-amps. As most of these devices are hybrid circuits they tend to be quite expensive. Nevertheless, if the application really calls for high gain, high precision, high CMRR and low noise, the additional expense can usually be justified. It is perhaps worth pointing out that amplifiers like the 725, which are called instrumentation operational amplifiers, are

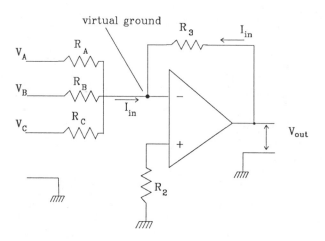

Fig 3.11 *A typical summing amplifier. Its operation is made possible by the existence of a virtual ground at the op-amp's inverting input.*

actually good quality op-amps. While they are excellent op-amps for use in laboratory electronics, they do not have the precision or other qualities of "true" instrumentation amplifiers - although they don't cost as much, either. Examples of some available instrumentation amplifiers are given in table 3.3.

Instrumentation amplifiers can be used in many of the circuits presented below, although such extravagance would not normally be justified. Furthermore, instrumentation amplifiers tend to have rather small bandwidths, being primarily designed for use with dc-producing transducers such as strain gauges, so their use in ac systems is somewhat limited. However, they are to be recommended as the amplifying components of high resolution analog input interfaces as described in chapter 7.

3.5.3 Summing amplifiers

The existence of a virtual ground at the inverting input of the inverting amplifier configuration op-amp, allows signals to be combined in a particularly straightforward manner. Figure 3.11 shows an inverting configuration amplifier with three independent signal voltages (V_A, V_B and V_C) applied to the inverting input via the three input resistors (R_A, R_B and R_C). Bearing in mind that the op-amp in the inverting configuration will do whatever it can to ensure that its output applied through the feedback resistor will maintain the inverting input at a virtual ground, we can see that the current through the feedback resistor (neglecting the op-amp bias current) must exactly balance the sum of the three currents flowing through the input resistors, ie.

$$I_{in} = -V_{out}/R_3 = V_A/R_A + V_B/R_B + V_C/R_C$$

Consequently the output voltage is a weighted sum of the three input voltages, ie.

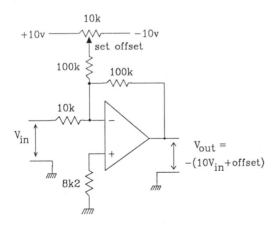

Fig 3.12 *Inverting amplifier with a wide range signal offset facility. The same arrangement may be used to null the op-amp's input offset voltage.*

$$V_{out} = - (V_A R_3 / R_A + V_B R_3 / R_B + V_C R_3 / R_C) \qquad (3.4)$$

This technique is known as "summing" and the inverting input of the op-amp in fig 3.11 is often referred to as the summing point. If $R_A = R_B = R_C = R_1$, then

$$V_{out} = - (V_A + V_B + V_C) * R_3 / R_1 \qquad (3.5)$$

and we have a technique for adding voltage signals together.

One useful application of the summing amplifier is in the provision of an offset, or the subtraction of a background level from a signal, before the balance of the signal is amplified or recorded. Figure 3.12 shows the circuit of a typical amplifier in which the input signal may be offset by up to 1 V (positive or negative) before the signal is amplified by the *10 amplification stage. Note that in this example the offset provision is for ± 1 V (before amplification), and not the ± 10 V applied to the ends of the potentiometer. This is because the input resistor in the offset input is 100 k, whereas that in the signal input is 10k. Thus the input signal voltage, V_{in}, is effectively multiplied by 10 while the applied offset voltage is multiplied by 1 (see eqn (3.4)). Lowering the input resistor in the offset input to 10 k would, of course, allow an offset of $\pm$ 10 V before amplification, while raising it to 1M would provide an effective offset range of $\pm$ 100 mV. If the potentiometer in fig 3.12 is a multi-turn preset resistor, then the circuit provides a universal method for offset nulling an inverting amplifier - although only suitable for the inverting configuration as it relies on the existence of a virtual ground at the summing point.

3.5.4 The voltage-to-current converter (Transconductance amplifier)

The high input impedance of op-amps makes it convenient for them to be regarded as voltage amplifiers, and their low output impedance means that they will maintain a required output voltage over a wide range of output currents. (Some op-amps are provided with in-

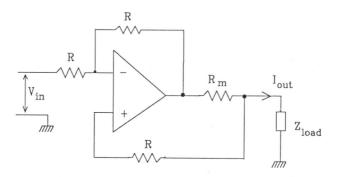

Fig 3.13 *One arrangement for the implementation of a transconductance amplifier, in which the output current is monitored by the voltage drop it produces across R_m.*

built protection circuits which shut off the output voltage if too much current is being drawn from the device.) However, there are occasions when a specific current signal is required under conditions in which it is not sufficient to rely on the load for that current remaining constant or at a constant reference potential. An example is provided by the need to generate a triangular waveform - such as that needed to modify the source velocity in Mossbauer spectrometry. A transconductance amplifier allows us to generate a current signal which can be driven into a varying load or a load operating with a changing reference voltage.

The current output of a transconductance amplifier is determined by a voltage at the input (hence the term voltage to current converter), and is given by

$$I_{out} = V_{in} \, g_m$$

where g_m is called the transconductance of the circuit and is in siemens (ie. reciprocal ohms).

The principle of one type of transconductance amplifier is illustrated in fig 3.13, where it can be seen that voltage feedback is applied to both inverting and non-inverting inputs from the ends of the resistor R_m. As usual the op-amp will adjust its output voltage in whatever way is necessary to ensure that its two inputs are at the same potential, and if no voltage signal is being applied (ie. $V_{in} = 0$), this can only occur when there is no current flowing through R_m - as only under those conditions are its two ends at the same potential. When the load (Z_{load}) is grounded on one side the zero current condition requires that the op-amp's output voltage is zero.

If an input signal, V_{in}, is now applied, the voltage at the inverting input is different from that at the non-inverting input, so the op-amp acts to oppose this change by applying the opposite potential difference across the inputs. This it does by passing a current, I_{out}, through R_m to produce the required potential difference, ie.

$$V_{in} = I_{out} \, R_m$$

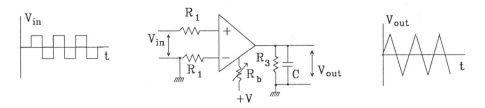

Fig 3.14 *A square-to-triangular waveform converter based on an operational transconductance amplifier, which produces an output current used to charge the capacitor C.*

So the output current from this configuration (which is independent of the load Z_{load}) is

$$I_{out} = V_{in} / R_m = V_{in} \, g_m$$

Transconductance amplifiers are available in single packages (although most of these are based on a different technique from that described above). An example is provided by the 3080 operational transconductance amplifier, which allows the transconductance to be varied by controlling a bias voltage or current at one of the IC's terminals. The 3080, which has a bandwidth of 2 MHz, may have its transconductance varied from essentially zero (with a bias current of zero) up to 9600 microsiemens (with a bias current of about 2 mA). A triangular waveform generator based on a 3080 device is illustrated in fig 3.14, where it may be noted that no external feedback element is used in this circuit. The waveform is generated by passing a constant current into a capacitor, the direction of this current alternating with the sign of a (voltage) square wave applied to the input of the 3080. R_3 is required if the circuit is driving a high impedance load, as this ensures that the output voltage varies symmetrically about ground. In the audio frequency range a value of about 100 k should be satisfactory.

If the load of a transconductance amplifier is a resistance, R_L, then the output voltage (ie. across the load) is

$$V_{out} = R_L \, I_{out} = V_{in} \, g_m \, R_L$$

so that the circuit behaves essentially as a voltage amplifier (in fact the 3080 offers quite a useful bandwidth for a voltage amplifier, considering the low cost of the device), but this voltage amplifier offers a gain which may be varied simply be varying the bias supply. Furthermore the gain can be changed from any particular value to zero and back - again by switching the bias supply - so that the signal being amplified may be gated, ie. passed according to the level of the (gating) signal on the amplifier's bias pin. Until recently these were valuable techniques to apply in the design of interfaces for microcomputers, although the improvements in the quality and speed of analog switches and programmable op-amps in the last few years have partly eclipsed the value of the transconductance amplifier in this role.

Operational transconductance amplifiers are relatively low-power devices capable of rather small output currents - typically a few microamperes - when compared with normal op-

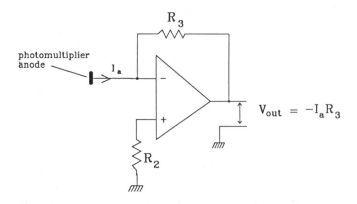

Fig 3.15 *A transresistance amplifier used to generate a voltage signal from the anode current of a photomultiplier tube.*

amp outputs, and require the addition of current amplifiers (eg. bipolar transistors) to generate higher currents. However, transconductance circuits based on op-amps and typified by fig 3.13 are also useful when substantial current signals are required or where higher voltage ranges than the normal ± 15 V need to be covered, as high output current or high voltage op-amp types may be used.

3.5.5 The current-to-voltage converter (Transresistance amplifier)

A transresistance amplifier produces a voltage output proportional to a current input. The applications of transresistance amplifiers may be more immediately apparent than those of transconductance circuits, because a number of important transducer devices (such as the photomultiplier tube) produce their output in the form of a current signal which is generally converted into a voltage signal before its level is shifted, amplified or recorded. A typical op-amp based transresistance amplifier is shown in fig 3.15. In this example the input is the current from a photomultiplier's anode, and it is worth noting that there is no other connection to the tube's anode. The tube is, of course, operating with a negative photocathode; this avoids the use of high voltage blocking capacitors in the signal lead and minimises interference from the high voltage supply and cabling. The anode is maintained at the required ground potential by the virtual ground at the inverting input of the op-amp, which in turn is present because the non-inverting input is grounded through R_2. To maintain the virtual ground the op-amp output voltage takes the value which causes the current in the feedback resistance, R_3, to be equal but opposite to that from the anode. Thus

$$V_{out} = -I_a R_3$$

The value of R_3 is chosen to produce a reasonable voltage output for the expected anode current; in our case we selected 10 V output for the maximum anticipated anode current of 0.1 mA, which required a feedback resistance of 100k and allowed our normal signal levels to be around 1 V. R_2 was chosen to equal R_3 as the latter is the only hardwired pathway for the input bias current to the inverting input.

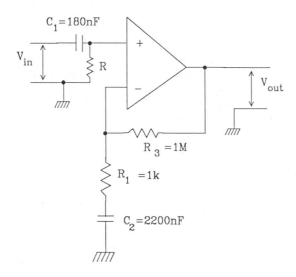

Fig 3.16 *A *100 ac voltage amplifier circuit which provides for a gain roll-off for frequencies below 10 Hz to minimise dc errors and drifts.*

3.6 Ac signal circuits

The circuits discussed above will handle ac voltage and current signals in addition to constant level signals, and often there is no need to worry about changing an amplifier's circuit when moving from a dc to an ac signal. However, there may be some advantages to handling an ac signal in a different way, and sometimes there is a need to handle ac signals unhindered by any superimposed dc level.

3.6.1 Ac amplifiers

Voltage and current amplifiers for ac-only signals can be assembled simply by including a capacitor in the input signal pathway of a dc amplifier. Such capacitors may be called dc-blocking capacitors or ac-coupling capacitors - depending on your point of view. Of course it is important to choose the value of the capacitance so that it provides a low impedance at the signal frequencies of interest (relative to the impedance of any gain-setting resistors). Figure 3.16 illustrates two points of interest in adapting straightforward amplifier circuits to handle ac-only signals. The particular example is for an audio frequency range, non-inverting configuration ac voltage amplifier, but the principles apply to all ac-only systems (of course, the choice of inverting or non-inverting in the ac-only case depends on the desired circuit characteristics such as stability and input impedance, as the inversion of an ac signal affects only its phase).

The first point is that if the dc pathway to either op-amp input is blocked by a capacitor then some alternative route for the input bias current must be provided. In our example this route is provided by the addition of the resistance R between the non-inverting input and ground. This resistor does load the input signal and, by strapping the non-inverting input to ground, ensures that a (time averaged) virtual ground exists at the inverting input. (This is useful if shifting of the mean signal level is required, as an offset may be applied at the non-inverting input.) The value of R may be selected in the normal way (ie. to be equal to the effective bias current pathway resistance to the inverting input), but note that R_1 does not provide a dc pathway to ground because of the blocking action of C_2. However, this consideration is not important if this circuit's output is ac coupled, because any dc offset errors will be blocked by the coupling capacitor. The input circuit made up of C_1 and R also forms a high-pass filter, with a corner frequency equal to $(2\pi C_1 R)^{-1}$, about 10 Hz with values of R = 100k and C_1 = 180 nF. Thus the input signal is attenuated at lower frequencies at the rate of 6 dB per octave.

The second point concerns the inclusion of capacitor C_2. One of the most frustrating forms of nuisance in a signal being recorded on, say, a chart recorder is the slow drift which results from an apparatus warming up over a period of several hours. This phenomenon can arise from temperature changes on a circuit board altering the input offset voltages of op-amps (see the input offset temperature coefficient, section 3.1), and is hard to control where the op-amps must amplify dc levels with a high gain. However, when ac-only signals are involved it is possible to roll-off the gain of an amplifier using the trick shown in fig 3.16. At zero frequency C_2 has an impedance of infinity, so 100% of the op-amp's output signal level is fed back to the inverting input and the amplifier has unity gain. At high frequencies the impedance of C_2 is much smaller than R_1, so the amount of feedback falls and the gain rises to $(R_1+R_3)/R_1$. With the values given in fig 3.16 the high frequency gain is 100. The corner frequency is that at which the impedance of C_2 = 10k, making the gain half of its higher frequency value, ie. at approximately 8 Hz in our example. Thus signals in the frequency range of interest (100 Hz - 10 kHz) are amplified 100-fold, while low frequency drifts and op-amp offsets are amplified by 1, or less for very low frequency noise on the input signal which has already been attenuated by $C_1 R$.

3.6.2 Ac to dc conversion

While it is convenient to handle ac signals for purposes of amplification, recording the value of such signals by chart recorder or computer requires that an ac - dc conversion takes place. In some cases it is adequate to rectify an ac signal by traditional methods - such as passing the signal through a diode followed by low-pass filtering to remove the residual ripple, as illustrated in fig 3.17. In this example the circuit's input signal should derive from a

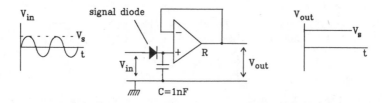

Fig 3.17 *The simplest technique for ac-dc conversion. It works over a wide range of frequencies but results in errors because of the diode drop.*

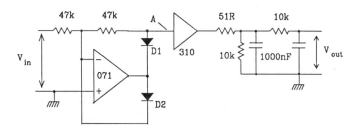

Fig 3.18 *A precision rectifier for ac-dc conversion of small signals without diode drop*
errors. Works for signal frequencies up to about 1 MHz with the op-amp shown.

fairly low impedance source to avoid excessive loading during the positive half cycles
resulting in errors. The output is buffered using a voltage follower to ensure that the effective
loading of C is small, so that the dc output records the peak ac input voltage less a "diode
drop" (typically 0.6 V for silicon diodes and 0.2 V for germanium ones). This circuit will
function up to high frequencies (eg. 50 MHz, if you take care to avoid stray capacitance
around the input) and its response time (ie. how rapidly it can follow changes in ac
amplitude) is given by CR^{-1} s, where R is the input impedance of the op-amp plus the
backwards resistance of the diode (ie. several M). With the values given the response time is
around 10 ms. If this is too slow for a particular application then a load resistor can be added
in parallel with C.

The main problem with this kind of circuit is that the diode characteristic is distinctly
non-linear, conduction through the diode producing a voltage drop across it. Although there
are special types of diodes which can help in this situation, an alternative technique is to
overcome the diode drop using an op-amp.

Figure 3.18 shows a typical "precision rectifier" circuit, capable of providing a dc output
voltage equal to the amplitude of the ac input voltage over the range of 1 mV - 5 V and 1 kHz
- 1 MHz (in this case the signal does not have to be an ac-only signal). For positive input
half-cycles D_1 conducts and D_2 is "off", so the circuit behaves as an inverting amplifier with
unity gain as far as the signal at the first stage output (point A) is concerned, although at the
op-amp output the voltage stays a diode drop below the input level. For negative input half-
cycles D_1 is off while D_2 conducts, forming an inverter through D_2 and allowing the op-amp's
output to stay a diode drop above ground (remember the inverting input is a virtual ground
when the non-inverting input is grounded). With the op-amp's output positive there is no
signal passage through D_1 to the buffer. The buffer is required in this type of circuit because
the output impedance of the first stage (ie. at point A) alternates between a low value (when
D_1 is conducting) and 47k (when D_1 is off and the virtual ground level (the inverting input) is
seen through the 47k resistor). The buffer presents a high input impedance to the half-wave
signal, so it doesn't load the signal at either impedance level, and provides a low impedance
output to drive the low-pass filter.

The circuit shown in fig 3.18 will handle small signals at frequencies up to about 1 MHz
when germaniun diodes are used. For silicon diodes the greater diode drop voltage changes

required as the signal passes through zero tends to restrict the application to frequencies below a few hundred kHz, because of slew rate limitations. For larger signal amplitudes slew rate limiting restricts the output amplitude in either case to a few tens of kHz.

3.7 Integrators

Important classes of circuits for instrumentation are those which enable a signal to be either differentiated or integrated. The differences between such circuits and those we have met already are actually quite small, although the effect produced on a signal may be large. An integrator is a circuit which produces an output voltage given by

$$V_{out} = \int V_{in}.dt$$

This function is achieved in principle by a circuit such as that shown in fig 3.19a. Here the op-amp's feedback element is a capacitor which charges up as the output voltage changes to pump a current equal to V_{in}/R_1 into the capacitor. This results in an output voltage given by

$$V_{out} = (CR_1)^{-1}\int V_{in}.dt$$

and CR_1 is termed the integration time constant. An example of the integrated output resulting from a constant V_{in} is also shown. Note that the slope of the output voltage is proportional to $(CR_1)^{-1}$, so that large integration times give a slower response.

An immediate problem arises with this circuit because its integrating action cannot be turned off without turning the power off. Two alternative circuits are useful in providing a practical integrator, the choice between them depending largely on the nature of the application and the timescale required for the integration. In the first circuit, fig 3.19b, a switch is provided to short out the capacitor and reset the integrator output to zero. This switch could be a manually operated switch, although this could result in the collection of interference signals which are likely to be picked up on wires between components and case-mounted switches. A better choice would be an FET or analog switch in which the signal remains close to the circuit board and the switch is operated by a simple control voltage applied to the FET gate or the IC's open/close pin. (Analog switches are IC electronic switches which have essentially infinite resistance ($>10^{12}$ R) when the open/close pin is at ground potential and a very low resistance (10 - 100R, depending on the switch type) when the open/close pin is at the positive supply voltage.)

This type of circuit is useful when integration is to be performed over periods of minutes, and can also be used for periods of seconds if the switch is operated automatically by the application of timed open/close signals. Typical applications are in the measurement of heat output (eg. calorimetry) or light output (eg. thermo-luminescence dosimetry) where the input voltage signal is provided by a suitable transducer. Incidentally the circuit can be used for current integration by omitting the input resistor. In any event the most likely difficulty with this kind of circuit arises from the inclusion of offset and bias currents in the integrated output, although the choice of an FET input op-amp helps to minimise this problem. For long-time integrations the use of a microcomputer to numerically integrate a recorded signal voltage is probably a more versatile and reliable alternative.

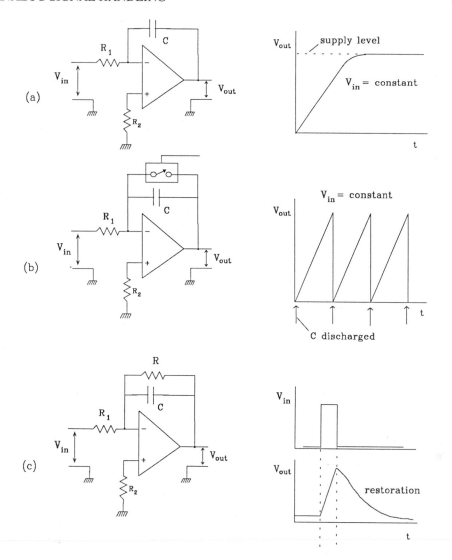

Fig 3.19 *The basic integrator circuit (a), which may be modified (b) to provide for a reset of the output to zero, or (c) to allow the output to be restored to zero with a fixed time constant given by CR.*

The second practical integration circuit, shown in fig 3.19c, avoids the output drift of the previous circuit by the addition of a restoring resistor, R, across the feedback capacitor. This allows the voltage across the capacitor to be restored slowly to zero, with a time constant given by CR and termed the restoration time constant. Although there is no reason why large value resistors (or other tricks to simulate them) should not be used, most applications of this technique are associated with relatively short time scale integrations, such as the integration of a current pulse from a nuclear particle detector or photomultiplier tube.

Without R_1 in the input line the circuit of fig 3.19c becomes a current integrator. As the integral of current is charge, this type of circuit is also called a charge sensitive amplifier, giving a (peak) output voltage proportional to the charge collected at the input during the integration time, ie.

$$V_{out} = C^{-1} \int I_{in}.dt = Q_{in} / C$$

Note the difference between this approach and that based on the use of a voltage amplifier operating on the voltage signal produced when the charge collected by a charge gathering transducer effects the electrode's potential through $DV = Q/C_T$ (section 2.1). The charge sensitive amplifier produces an output which is largely independent of the transducer's capacitance C_T. In this case it is the maximum value of the integrated output signal which provides the required information, and this is proportional to C^{-1}. The restoration time constant may be chosen to be as large as possible while avoiding the overlap of adjacent signals.

3.8 Differentiators

The basic differentiator circuit is shown in fig 3.20a and produces an output voltage given by

$$V_{out} = -RC (dV_{in} / dt)$$

In this case the feedback element is a resistance and the input voltage signal is coupled to the op-amp's input by a capacitor. An example of a differentiated output produced by a pulse input is shown in the figure, although the form of the output would depend on the pulse duration and the component values used in the circuit. The time constant given by the product RC is called the differentiation time constant, and determines the magnitude of the output voltage for a given rate of change of input signal. As the feedback element in the differentiator provides a dc pathway for the op-amp's bias current, differentiators do not suffer from the drift problems of simple integrators, although the basic circuit shown does produce an output which can be distorted by contributions from high frequency noise and interference. Furthermore differentiators which are not frequency limited are prone to high frequency oscillation. The addition of frequency limiting components, which cause the gain to roll off at high frequencies, makes the circuit, shown in fig 3.20b, look similar to that of the integrator in fig 3.19c. The resistor R_1 in the input line limits the gain of the circuit by providing an upper limit of R/R_1 at high frequencies (where the impedance of C has become small), while the capacitor C_2 ensures that at high frequencies the circuit behaves as an integrator with an integration time given by R_1C_2 s.

Finally it is worth bearing in mind that pulses form a class of signals which inconveniently fall between the realms of dc and ac. They often require handling circuits which have high bandwidths, so they sound like good candidates for amplification by ac circuits with capacitors to pass the signal from one stage to the next. However, passing a unipolar pulse through a capacitor to a load resistor results in at least partial differentiation of the signal. For this reason it may be better to differentiate the input pulse in a precise manner and produce a bipolar signal pulse in the first place, as these pseudo ac signals are less

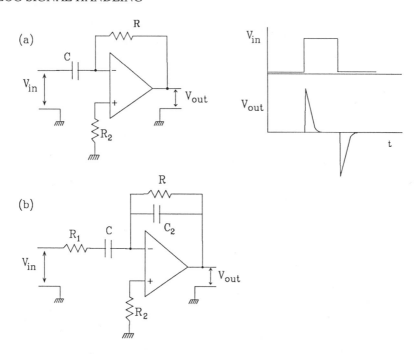

Fig 3.20 *(a) The basic differentiator circuit. (b) A differentiator with a high frequency roll-off to limit high frequency noise and spurious oscillation. This circuit forms the basis of a pulse shaping amplifier.*

susceptible to distortion through interactions with stray circuit impedances than are unipolar signals. Also bipolar signals could then be handled by ac-coupled circuits, which tend to have fewer stability and offset problems than their dc-coupled counterparts. On the other hand ac-coupled circuits do have more complex frequency dependencies of their transfer functions, so that if a wide range of pulse shapes needs to be accommodated then the choice is more likely to fall to the dc-coupled system.

3.9 Pulse amplifiers

Pulse amplifiers are essentially refined versions of the type of circuit shown in fig 3.20b. Many commercial pulse amplifiers, particularly those designed for nuclear instrumentation, are provided with switch selectable time constants for both differentiation and integration. The choice of these time constants depends on the information desired from the pulses, and the effects of various combinations of time constants on the output signals generated from a standard input "step" (an abrupt change in input voltage usually adopted as an extreme model of the long tail pulse of fig 2.3d) are illustrated in fig 3.21. If the magnitude of the pulse height is desired (eg. for pulse height analysis) then this may best be measured using a short integration time constant (comparable to the rise time of the incoming signal), as this allows the most accurate pulse height conversion. Similarly if a long differentiation time constant is chosen then the peak of the output signal decays more slowly and is therefore easier to

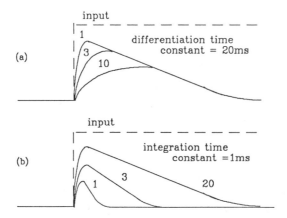

Fig 3.21 *Outputs of a pulse amplifier for an input voltage step, showing the effects of varying (a) the integration time constant while the differentiation time is fixed, and (b) vice versa. The value of the varying time constant is shown beside each trace.*

measure accurately. However, against this one must note that with a long differentiation time constant the output pulses may overlap one another, making the accurate measurement of signal pulse heights less reliable because of pulse "pile up".

On the other hand, if fast pulse counting is the objective then preserving the pulse height is not as important as ensuring that the pulses are not missed by accidental overlapping, so short differentiation time constants are attractive, with the integration time constant chosen to maximise the pulse height. Unfortunately short integration time constants will allow fast noise pulses to be recorded. Sadly there is no infallible guide to how one should select these time constants, and the choice will depend on the application and the shapes of signal and noise pulses. A useful starting point is usually to match the time constants to the rise and fall times of the input signals. However, it is important to realise that the selection of different time constants does have a profound effect on the shape of the output signal. In fact, by carefully selecting the time constants it is possible to discriminate against some shapes of input pulses while amplifying others, and this forms the basis of one technique of distinguishing between different types of ionising radiation detected by nuclear radiation detectors.

3.10 Filters

Filters are circuits which have transfer functions which allow the passage of signals of one frequency, or range of frequencies, while blocking those of other frequencies. Like most of the topics covered in this text, the subject of filter design is very large and we can do no more that examine briefly one or two examples of important types of filters. We will consider examples of active filter circuits (ie. those containing at least one powered device capable of amplification), although filter circuits can be constructed using only passive (unpowered) components if a loss in signal amplitude is acceptable.

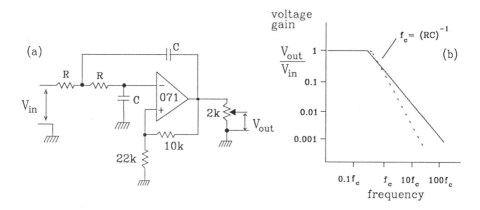

Fig 3.22 *A two pole low pass filter (a) giving a frequency response (b) with a corner frequency, f_c, determined by the two RC networks. The broken line shows the response obtained by connecting two filters in series.*

The first example is a low pass filter - a circuit which passes signals of frequencies lower than the circuit's corner frequency, f_c, while attenuating signals of higher frequencies. The circuit is shown in fig 3.22a and its response to ac signals as a function of frequency is in fig 3.22b. The filtering action is brought about by the RC networks in the op-amp's non-inverting input line and feedback loop - two RC units, and it is called a two pole filter. Because the two RC networks have the same component values the response of this filter is known as a Butterworth characteristic, and above the corner frequency of RC^{-1} the gain declines at 12dB per octave (in terms of voltage gain that is a factor of about 4 for each doubling of frequency). By selecting the two RC sections to have different component values, other rates of gain decline may be chosen, Chebyshev (faster decline) and Bessel (slower decline) being commonly used filter characteristics, although having a more variable low frequency gain than the Butterworth type.

The preset variable resistor providing the output from the filter circuit of fig 3.22a may be adjusted to produce an overall gain of unity for low frequency signals (without that facility the voltage gain is actually about 1.6 when the Rs are 10 k) and these circuits may be cascaded to produce a sharper cut-off above the corner frequency. Low pass filters are particularly useful for the attenuation of high frequency noise accompanying dc or low frequency signals produced by laboratory transducers. A corner frequency of less than a few hertz enables dc signals to be cleaned-up by the removal of interference, such as that derived from the 50 or 60 Hz line frequency.

Interchanging the Rs and Cs of fig 3.22a produces a high-pass filter of the type shown in fig 3.23a and this has a response of the form shown in fig 3.23b, allowing the passage of signals with a frequency above the corner frequency, while attenuating dc and low frequency signals. Again different filter characteristics can be obtained by selection of the component values, although of course in this case an upper frequency limit to the transmitted signal is imposed by the bandwidth and slew-rate of the op-amp used. Combining low and high pass

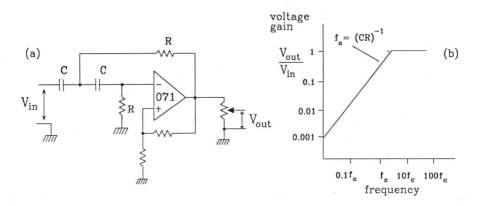

Fig 3.23 *A two pole high pass filter (a) and its frequency response (b).*

filters with overlapping pass frequency ranges gives the band-pass filter, which passes a signal within a given frequency range (the pass band) while attenuating signals outside that range (both lower and higher frequencies). In much the same way summing the outputs from non-overlapping low and high pass filters can be used to generate band-stop filters. However, it is not so easy to cascade band-pass and band-stop filters because of the sensitivity of the band frequencies to the precise values of the components in the RC networks.

While the filter circuits we have considered have been based on conventional op-amps, each type of filter characteristic may be obtained from purpose designed ICs known as state-variable filters. These enable high quality filters of each type to be constructed with a minimum of difficulty and only a small number of additional components. Furthermore, the manufacturers usually supply detailed tables of component values required for the different filter characteristics and for a wide range of frequencies extending from Hz to MHz. A typical low cost and versatile device is the MF10C, which can be used with signal frequencies up to about 200kHz, although with corner frequencies only as high as 20kHz. This device may be configured for high, low or band pass, and for Butterworth, Bessel, Cauer and Chebychev characteristics.

3.11 High voltage and high power control circuits

As we shall see in chapter 7, it is a relatively straightforward matter to generate low voltage analog signals with the aid of a computer. However, in many laboratory systems it is desirable to control much higher voltages or currents than can be handled by the amplifier devices we have met so far. Consequently we conclude this chapter with some examples of circuits which allow a low power source to control the value of a much higher voltage or current.

High voltages are commonly encountered as the bias potentials for photomultiplier and electron multiplier tubes and the gas ionisation detectors for ionising radiation. Generally it is necessary for the bias potential to be set to a specified value and for it to remain at that value while the transducer is in operation. Actually generating a high voltage from a low one is not

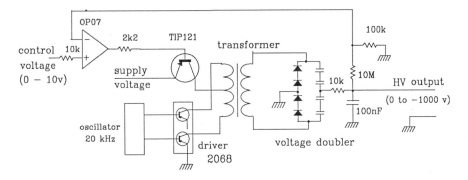

Fig 3.24 *An outline circuit for a controllable high voltage bias supply which produces an output voltage which is 100 times the control voltage applied to the op-amp.*

particularly difficult. One starts with a low voltage ac signal applied to the primary winding of a "step up" transformer, and rectifies the high voltage ac developed across the transformer's secondary winding. Furthermore it is common practice to use a series of high voltage diodes to double the voltage generated across the secondary. However, on their own these techniques generate a voltage which varies substantially with the current drawn by the load - ie the transducer being biased by the high voltage. Consequently some form of feedback circuit is required to ensure that, within the circuit's capabilities, the output voltage is maintained constant whatever current is drawn by the load.

A circuit which has been used in a number of the instruments in our laboratory is shown in fig 3.24. A 20 kHz oscillator, producing a square waveform between 0 and the supply voltage (generally 15 V), is used to alternate the direction of current flow through the primary winding of the transformer. The magnitude of the current flow is controlled by using the difference between a portion of the circuit's HV output (divided down by the 10M and 100k resistors) and a low value control voltage, to bias the base of the TIP121 power transistor. The transformer, which has a turns ratio of about 20:1, produces a high voltage ac output which is rectified and doubled by the arrangement of diodes and capacitors connected as a "voltage doubler." The 10k resistor in the output line is present to limit the current in the event of an accidental short circuit on the output, and, together with the 100 nF capacitor, forms a low-pass filter with a time constant of $(10^{-8}*10^4 =) \ 10^{-4}$ s, enough to reduce the 20 kHz ripple on the output voltage passed through the rectifier. As the divided-down HV value returned to the difference amplifier is 1% of the actual high voltage, a control voltage of 5 volts limits the transformer current to whatever is required to generate an output of 500 V, irrespective of small changes in the current being drawn from this high voltage output. In this example the source of the control voltage is unspecified, although in our laboratory instruments it is generally provided by a computer generated reference voltage, allowing the computer to control the high voltage level with a high degree of precision. With a 2068 driver, the circuit shown can provide up to 1 mA at 1000 V to bias a photomultiplier tube.

The circuits shown in fig 3.25 may be used for controlling substantial current flows at low voltage levels, the kind of power required for dc motors, small heaters or light sources.

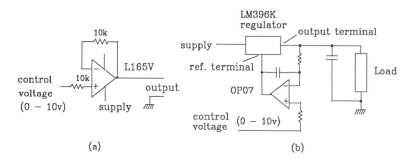

Fig 3.25 *Two circuits which may be used as the basis for controllable low voltage, high current supplies. The circuit in (a) is based on a high power op-amp, while that in (b) uses a conventional (low-power) op-amp to control an adjustable three terminal voltage regulator.*

Figure 3.25a shows a power op-amp, similar in principle to any other op-amp, although generally manufactured in a package suitable for heat sinking and capable of supplying output currents one or two orders of magnitude larger than those available from conventional op-amps. The L165V for example, can deliver up to 3 A and its output may range from +12 to -12 V. We have used this kind of circuit to drive small dc motors and to power sample heaters (actually constructed from soldering iron elements).

The circuit shown in fig 3.25b uses a conventional low-power op-amp to provide a reference voltage for a three terminal adjustable voltage regulator. These regulators typically generate an output voltage which is a fixed amount greater than the voltage present at their reference terminal; the LM396K, for example, produces an output which is 1.25 V greater than its reference voltage, and can pass up to 10 A when suitably mounted. The circuit shown will generate a regulated output voltage in the range 1.25 - 11.25 V, independent of the load resistance, and supply several amperes. The output voltage is determined by the low power control voltage applied to the non-inverting input of the op-amp. The inverting input of the op-amp senses the voltage applied to the load (and for high current systems should be connected as near to the load as possible to prevent errors through voltage drops in the current-carrying conductors). The output of the op-amp will do whatever is necessary to ensure that the non-inverting input is at the same potential as the inverting input - in this case by generating an output which is 1.25 V greater than that at the non-inverting input so that the regulator produces an output voltage equal to that at the non-inverting input. We have used variants of these circuits for controlling the intensity of light sources in colorimetric detectors used for flow analysis instruments.

In all of these circuits (figs 3.24 and 3.25) it is important to note that the power limitation is governed not by the current flowing through the high power component (the TIP121, the L165 and the LM396 repectively), but by the temperature produced by the power dissipated in that component. Thus the LM396K can handle currents up to 10 A, but only if that current flow does not cause the temperature within the device to exceed its rated maximum (this particular component will shut down at 110°C; other devices just self-destruct). The temperature within any power component is governed by the power dissipated within the component (ie how quickly it is heating up) and the efficiency of heat conduction away from the component (ie how quickly it is being cooled down). The second factor may be adjusted by carefully designed heat sinking of power components, mounting the device on a matt black thermal conductor equipped with large surface area radiation panels. (For a quantitative discussion of heat sinking the reader is referred to "The Art of Electronics", see bibliography). However, it is also essential to understand that the power dissipated within the device depends not only on the magnitude of the current flowing, but also on the voltage drop across it. Thus if the circuit in fig 3.25b is adjusted so that it supplies an output current of 10 A at a voltage of 10 V, then the power dissipated in the LM396K is given by the product of the 10 A current flowing through it and the voltage dropped by lowering the supply voltage from whatever it is to 10 V. If the supply voltage is 20 V then 100 W is being dissipated within the LM396K, but if the supply voltage is only 12.5 V (the design of the LM396K is such that at least 2.5 V will be dropped across its terminals) then only 25 W is being dissipated in heating up the device. The heat sinking requirements will obviously be rather different in these two examples (in fact the first would be virtually impossible to accommodate).

CHAPTER 4

THE ELEMENTS OF DIGITAL SIGNAL HANDLING

By comparison with analog signals we should find digital signals quite straightforward, because they lack the variety of forms which plague the analog world. A digital signal can have only one of two possible values on any signal carrying conductor at any given time. The two possible values are known as logic levels, and for the purposes of generalising are referred to as "high" and "low" levels (more commonly abbreviated to hi and lo). The simplest electronic components designed specifically for handling these signals are called logic gates, and there are hundreds of gates on the market - each designed to produce one or more outputs whose level depends on the level applied to one or more inputs. To utilise these circuits it is necessary to define electrical signal values to represent the levels of hi and lo. One of the most widely used conventions requires the level hi to be represented by a voltage of +5 V, and the level lo by 0 V, both being relative to ground. Later in this chapter we shall examine this convention a little more closely and consider one or two alternatives.

4.1 Logic gates

Figure 4.1 illustrates the schematic representations (the symbols) used for two types of logic gate. In fig 4.1a the two input AND gate is shown. The output level from this gate is hi when the levels applied to input a and input b are both hi. If either or both of the inputs are lo, then the output is lo. The two input OR gate is illustrated in fig 4.1b. In this case the output of the gate is hi if either input a or input b is hi. If both inputs are lo then the output is lo. If both inputs are hi then the situation may be regarded as meeting the requirements for a

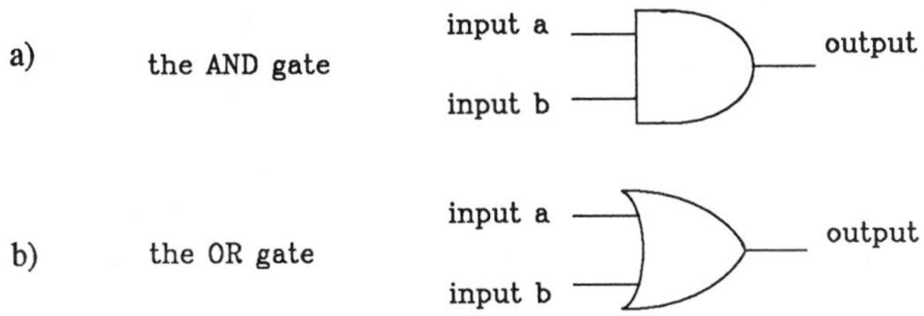

a)   the AND gate

input a ———
                    output
input b ———

b)   the OR gate

input a ———
                    output
input b ———

Fig 4.1 *Symbols used in circuit diagrams to represent AND and OR logic gates. The devices also require a power supply and a ground connection, but these are not usually shown.*

hi output, and this gate is more properly termed an inclusive-OR gate (ie. its operation is inclusive of the two inputs hi condition for hi output). An alternative gate, called an exclusive-OR (EOR) gate, treats the two input hi condition as not meeting the requirements for output hi, and gives a lo output in that situation.

Describing the action of gates is greatly simplified by the use of a logic symbolism, and this has the additional advantage of introducing a digital significance of the signal levels. The symbolism allows us to use 1 and 0 or "true" and "false" (after Boole) for the signal levels hi and lo. By far the most widely used logic system is that in which a 1 represents the hi level and a 0 represents the lo level, and this is the logic system used by manufacturers in naming the functions of their gates. Naturally the alternative is also used, and in the context of this book this alternative appears in a number of important applications (for example, the GPIB in chapter 8). The system with hi = 1 is called positive logic, or sometimes hi=1 or high-true logic. We shall use this system wherever possible. The alternative system which uses hi = 0 is termed the negative logic system. The difference between the two systems is of major importance, as may be seen by writing out the symbolic representation of the operation of the AND gate using both positive and negative logic as shown in table 4.1. This type of representation is known as a truth-table and is the principal description of the operation of a gate in manufacturers' data sheets. The negative logic output for this gate symbolises a different function from the positive logic output, as in one case different inputs give the same output as two 0s, and in the other case different inputs give the same output as two 1s. In fact the function achieved in the negative logic section of table 4.1 is not the AND function (output 1 only when both inputs 1) but the inclusive-OR function (ouput 1 when either input 1). (Of course the gate doesn't know the difference, it always produces a hi output only when both inputs are hi.)

Two other important gate functions are the EQ and the NOT functions, although the gates which implement these functions alone are commonly called buffer and inverter gates respectively. The gate symbols and truth-tables for these functions are given in fig 4.2. Note that the difference between the two symbols is the presence of the small circle on the output of the NOT function gate. The NOT function is most commonly encountered in combination with some other function, such as AND or OR, and gates which implement the combined

Table 4.1   Operation table for a two-input AND gate and associated truth tables

electrical operation			positive logic			negative logic		
a	b	output	a	b	output	a	b	output
lo	lo	lo	0	0	0	1	1	1
lo	hi	lo	0	1	0	1	0	1
hi	lo	lo	1	0	0	0	1	1
hi	hi	hi	1	1	1	0	0	0

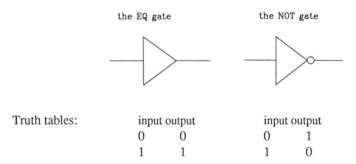

the EQ gate   the NOT gate

Truth tables:	input	output		input	output
	0	0		0	1
	1	1		1	0

Fig 4.2 Symbols and truth tables for the gates providing the EQ (buffer)
and NOT (inverter) functions.

functions are probably the most widely used of all logic gates. The symbols and truth table
for the NOT AND combination, abbreviated to NAND, and the NOT OR combination,
abbreviated NOR, are shown in fig 4.3. Note again the presence of the small circle on the
output lines, indicating the addition of a NOT function to the function symbolised for the
gate.

The real power of circuits made from logic gates lies in the ways in which the individual
gates may be combined to produce ouputs whose levels depend on the levels of several
different inputs. We can demonstrate this in a small way by examining the circuit shown in
fig 4.4. This is designed to react in a precisely defined way to the action of three switches:
one used for sensing whether a door is open or closed; another operated by a relay connected
to the power-on circuit of an X-ray generator; and a third used as a reset switch. In operation
the circuit could be used to activate an alarm (which could be a bell, or a light or a relay to
turn the power off) if the door of a safety enclosure of an X-ray diffractometer is opened
while the power is turned on to the X-ray generator. The alarm does not turn off again even if
the door is closed or the X-ray power turned off, unless a reset switch is operated.

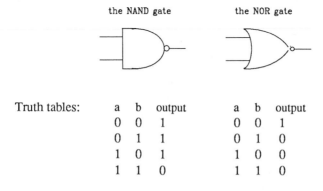

the NAND gate   the NOR gate

Truth tables:	a	b	output		a	b	output
	0	0	1		0	0	1
	0	1	1		0	1	0
	1	0	1		1	0	0
	1	1	0		1	1	0

Fig 4.3 Symbols and truth tables (positive logic) for the gates
providing the two input NAND and NOR functions.

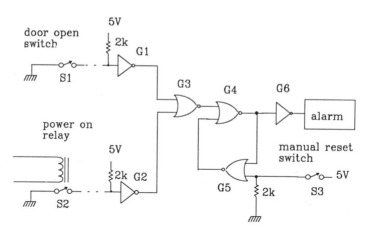

Fig 4.4 *Example of a simple logic system constructed from gates. S1 is closed when the door is closed, and S2 is closed when the generator power is off. The reset switch (S3) would be a momentary action, normally-open switch.*

Furthermore the alarm operates in the event of a breakage in the wires transmitting the signal from either the door switch or the relay switch. (Note: this circuit is provided as an example of the use of logic gates and not as an example of safety systems. The author does not claim to be qualified in the design of safety systems.)

For simplicity we shall assume that the door switch provides a 0 signal (ie. 0 V) when the door is closed, and that the relay provides a 0 signal when the generator power is off. If either switch opens, or the relevant wire is broken, then the input to the gates, G1 or G2, is taken to a 1 (eg. +5 V) by the 2 k "pull up" resistors. Gates G1 and G2 are inverters and output a value which is the complement of the input signal. G3 is a NOR gate and provides an output of 1 only when both inputs are 0s, ie. when either the power to the generator is on and the door is open, or when one of those conditions has been met and there is a fault in the other input channel resulting in a 0 level at G3's input. Thus the output of G3 is a 0 under normal conditions but becomes a 1 when the alarm is required to operate.

If the alarm was connected at the output of G3 then it would operate under the required condition (door open and power on), but it would stop operating if the door was closed again. We choose to complicate matters a little further (and to illustrate a useful technique) by insisting that the alarm, once activated, continues to operate until a reset switch is toggled. Gate G4 is a NOR gate accepting one input from the sensing circuit (ie. G3) and another from the output of a second NOR gate, G5, which has one of its inputs connected to G4's output and the other connected to a normally-grounded line from the reset switch. Consider first the situation in which G3's output is a 0 (ie. no alarm) and G5's output is a 0. Both inputs to G4 are 0s, so G4's output is a 1. Thus one input to G5 is a 1 and the other is a 0 (being held at ground by the 2k resistor until the reset switch is operated). Under these conditions the output of G5 is a 0, so providing G4 with its second 0 input.

On the other hand when an alarm condition is sensed the output of G3 becomes a 1, so the output of G4 becomes a 0, the output of G5 becomes a 1, and this ensures that the output of G4 remains a 0 even when the output of G3 reverts to a 0 (ie. the alarm condition is removed). Only when the normally grounded input to G5 is switched to a 1 (the 5 V signal connected to the other pole of the switch) does the output of G5 become a 0, allowing the two 0 inputs to G4 to produce a 1 from G4's output again. Finally the output of G4 is connected to an inverter, G6, which produces the 5 V output to operate the alarm only when its input is a 0 (at this stage we assume that the alarm operates on receipt of a 5 V signal level, and we refrain from considering the amount of current it may need).

It is worth spending some time ensuring that the operation of the circuit in fig 4.4 is understood, as it illustrates not only the connection of logic gates to produce a circuit which behaves in a more complex, but still logical, manner, but also the important technique of the "hold until reset" circuit. As an exercise you may like to try to improve the circuit, by reducing the number of gates used.

4.2 TTL families

Before we move on to consider more sophisticated logic functions we will examine briefly some of the commercial implementations of the simple gates we have met so far. All commercial gates are available as integrated circuits - at this level they are called small scale integration, SSI, circuits - packaged in DIL packages having 14 connections. As this is a larger number of connections than required for the one and two input gates we have considered above, it is not surprising that most of the 14 pin DIL packages actually contain several gates, typically four two-input NAND gates etc. Thus in fig 4.4 the gates G3 - G5 could all be in the same package, as could G1, G2 and G6. If you had replaced G1-G3 with a single AND gate the circuit would actually have required more packages than it does as shown. (What could you do to lower the package count?)

There are several different "families" of gates available, although we shall consider only two classes of these. The first class is that based on circuitry known as transistor-transistor logic (TTL) of which the oldest family is that known as the 74 family. Some examples of 74 family packages are listed in table 4.2, showing the basic numbering system which has been adopted by all major manufacturers and inherited by the newer families we shall discuss below. All the TTL families operate with supply voltages of +5 V and ground, but there are other characteristics which are different for the different families. There are four major factors which determine the suitability of a family for any particular application:

1. The signal levels used to represent the hi and lo levels.
2. The speed with which gates can change their signal levels.
3. The number of inputs which may be directly connected to a single output (known as the fan-out).
4. The power required to operate each gate.

For the 74 family (and in fact for all the TTL families) the signal levels are voltage ranges, with fairly stringent current requirements which cannot be neglected. The hi level for the 74 family is a positive voltage in the range 2.4 - 5 V, although 3.5 V is a typical value

Table 4.2 Examples of 74 series packages of gates

code number	no. of devices per package	type of device
7400	4	2-input NAND
7401	4	2-input NAND oc
7404	6	inverters
7406	6	inverters oc
7409	4	2-input AND oc
7410	3	3-input NAND
7420	2	4-input NAND
7430	1	8-input NAND
7402	4	2-input NOR
7408	4	2-input AND
7432	4	2-input OR
7486	4	2-input EOR

found for the hi output of a gate. This output is not intended to be a source of current for driving low impedance devices such as indicator lamps, and in fact the current available from hi outputs is quite small. The lo level is a voltage between 0 and 0.8 V, although gates providing a lo output produce a voltage which is within a few tenths of a volt of ground. One important characteristic of the lo state is that as an output level it is capable of sinking up to about 16 mA of current (ie it will absorb up to 16 mA from a positive potential source) in an attempt to achieve the required output voltage. (So in fig 4.4 it would have been better to operate the alarm with a 0 output rather than a 1.) To produce a lo level at the input of a 74 family device we need to be able to sink about 1.6 mA to ground (whether this is achieved by the output of a previous gate or by some other means). As a result of the requirements of the 74 family lo level, the fan-out of the 74 family is 10, ie. we may connect up to 10 gate inputs to a single gate output. Of course two outputs should never be directly connected together, unless they are special gates known as "open collector output" gates. These require a resistor to be placed between the output and a hi point. When a lo level is to be produced the open collector output drops to ground, sinking current through the resistor, but otherwise an effective hi output level is maintained through the resistor by the hi point and the gate output draws no current. Typically a resistor of 2-5 k is used for "pulling up" an open collector output, so that it can provide current for other TTL inputs, although if the output is not driving TTL a higher value resistor may be appropriate.

One important point associated with the outputs of the 74 family devices is that they generate a large current spike on the power supply line whenever a gate's output changes level. Thus a good quality bypass capacitor is required as close as possible to the power supply connection to every 74 family package on a circuit board. Omission of this small detail can give rise to the most chaotic circuit behaviour, particularly when other types of IC are present on the same circuit board.

Fig 4.5 *Example of a "glitch" on a logic line. Glitches can cause fast TTL systems to
misbehave, and their avoidance requires careful and neat layout of circuit boards
and adequate decoupling of power lines close to each TTL device.*

The speed with which a TTL gate can operate is most usefully measured by the gate
propagation time, which is the time taken for a change in one input level to bring about a
change in an output level (the level change itself occurs much more rapidly). For the 74
family the gate propagation time is approximately 10 ns, so that circuits built up from 74
family devices can handle signal levels changing at regular rates up to about 35 MHz,
although the normal design maximum is best restricted to 25 MHz. This is a usefully high
speed and, together with the simplicity with which devices may be connected to one another,
accounted for the early popularity of the 74 series. The price paid for this high switching
speed is the power consumed by the gates, which is about 10 mW per gate. This doesn't
sound like a lot of power, but it must be remembered that each package can house several
gates, and that complex systems may use hundreds or thousands of gates, so that the power
drawn by a large system based on the 74 family can become significant. In practice 74 family
packages operate warm to the touch, and when a large number of packages are housed in a
small instrument case the resulting heat generation can cause problems for other devices,
such as op-amps.

In practice the very high switching speeds of TTL gates and the large current spikes that
gate operation generates on the power supply lines (one of which is ground), can be the cause
of some difficulty. One of the secrets of success with TTL is to keep power line and signal
line inductances as small as possible to minimise the generation of glitches. (Glitches are
extremely fast pulses produced in one gate during the operation of another gate. It is
sometimes difficult to see these glitches on an oscilloscope unless the 'scope is both fast and
provides time-base control that allows you to examine both ends of a main pulse. A typical
glitch is shown in fig 4.5, and while they may be difficult to see, TTL gates manage to
respond to them all right!) Unfortunately keeping inductances low is incompatible with
bread-boarding a circuit with lots of interconnecting wires. As a result the development and
testing of TTL designs is not always a joy, and it usually causes far less aggravation to build
a high speed system with a carefully planned printed circuit board in the first place.

Over the years other families of logic gates have been developed from the basic TTL
designs. A high power family was produced to allow greater switching speeds at the expense
of greater power dissipation, and a low power family offered lower power dissipation
although operating at lower switching speeds. The addition of Schottky diodes across most of
the transistors which make up normal TTL gates results in a fairly dramatic increase in speed
- although again at the expense of higher power dissipation - and provides the Schottky TTL

Table 4.3 TTL families and principal characteristics

family name	typical code	switching speed /MHz	power per gate /mW
Standard	7400	35	10
High power	74H00	50	22
Low power	74L00	3	1
Schottky	74S00	125	19
Low power Schottky	74LS00	45	2
Advanced low power Schottky	74ALS00	90	1
Fast	74F00	150	6

family. Increasing the impedance levels throughout the structure of each gate slows the devices down but enables them to operate with lower currents. This technique coupled with the use of Schottky diodes has resulted in the development of the Low-power Schottky family, which is only slightly faster than the original 74 family but offers the advantage of much lower power requirements (about 1/5th of that of the 74 family). In the late 1970s an Advanced Low-power Schottky family was introduced, offering double the speed and half the power demand of its predecessor. A Fast family is also available and intended primarily for high performance circuits. The Fast family gates operate at about twice the speed of the Advanced low-power Schottky gates at a cost of significantly increased power consumption - although still substantially lower than the Standard family.

Each family of devices can be identified by its sensible identification code. Examples of these codes and the speed and power characteristics of each of the TTL families are summarised in table 4.3. All the families have a fanout of ten with gates of the same family, but care is needed if gates of different families are to be connected together because the lower power families are capable of sinking less current than the others. Data sheets should be carefully studied if members of different families are to be connected with fanouts of greater than two. Manufacturers usually recommend that the inputs of any unused gates should be connected to the +5 V supply, to prevent spurious level changes of the unused output from contributing spikes to the power rails. On no account should unused inputs be grounded - this only causes the package to use more power than is necessary. Devices which belong to the 74 families are specified for operation over the (package) temperature range of 0 - 70°C. The same gates specified for operation over the military temperature range (-55 to +125°C) are identified as the 54 families.

4.3 C-MOS families

The other class of logic gate families of major importance is that based on Complementary Metal Oxide - Silicon (C-MOS) technology. The modern families available in this class are all low-cost, with low-power requirements and are considerably more pleasant to use than the TTL families. C-MOS gates have very high input impedances in both hi and lo states (they use fet-type inputs), so that virtually no currents pass between gates and the power supply spikes characteristic of TTL level changes are virtually non-existent. As a result the chances of a circuit misbehaving through pick-up of spurious signals from other gate connections or power rails are very much smaller than with TTL. Some of the C-MOS families have been designed as a pin-for-pin replacement for the TTL families. One is identified with the coding 74C, and the latest arrivals are the high speed families identified as 74HC and 74HCT. As these latter have the speed of 74LS TTL together with the advantages of C-MOS they are undoubtedly destined to become very popular. (The 74HCT family is fully compatible with the 74LS family, but requires a thousand-fold less power for operation!) In common with other 74 families, the 74HC and 74HCT families operate with a supply of +5 V, while the difference between them is in the definition of the hi and lo logic levels - the 74HCT devices adopting the same definitions as we met for TTL, and the 74HC devices utilising the "normal" C-MOS definitions described below. In this writer's view these are the families of choice for most logic circuits required for in-house computer interfaces, particularly as they are generally less expensive than their older relations.

Having said that, there remains another group of C-MOS families called the 4000 families, and, as at least one of these remains very widely used and offers the attraction of being particularly tolerant to rather hastily constructed "bread-board" circuitry, it is appropriate for us to discuss this family in some detail.

The original 4000 family was developed by RCA, but was soon replaced by a modified form called the A-series, with its packages designated 4001A etc. This series had a few minor problems, such as gates which did not work, and packages which burnt out if the power supplies were turned on in the wrong order! Most of the A-series has now been replaced by the B-series (the B is for "buffered" apparently) and these are the C-MOS family of choice for most systems. Some applications (such as level changing) call for unbuffered gates and the improved versions of these, which otherwise meet the B specifications, are identified as UB devices. Examples of some of the B-series gates are given in table 4.4.

The 4000B series gates use very little power and most will operate with power supply levels anywhere within the range of +3 to +15 V. (Note that other C-MOS families such as 74HC require +5 V.) Thus the 4000B series can be used easily in battery powered instruments, although it must always be borne in mind that if a gate is used to supply current to something, such as a resistive load or a TTL family logic gate, then the power to drive that load comes from the gate's power supply, and can easily be hundreds of times greater than the power needed to operate the C-MOS gates alone. The devices are capable of only low current outputs, typically sourcing up to 0.2 mA in the hi state and sinking up to 0.5 mA in the lo state when operating with 5 V supplies, and about three times these values at 10 V. (Higher currents are available from the 4049 and 4050B devices, see section 4.4.) Furthermore ' the low gate-power advantage is only realisable when all the gate inputs are connected to either a hi or a low level. Unused inputs on a package must not be allowed to float - a rule which is even more important for C-MOS gates than for TTL. The best course is to connect any unused inputs to the positive supply line.

One consequence of the lower gate power demand is that the C-MOS gates have slower switching speeds than the faster of the TTL families (although the 74HC family seems to

Table 4.4 Examples of C-MOS 4000B series devices

code number	no. of devices per package	type of device
4011B	4	2-input NAND
4023B	3	3-input NAND
4012B	2	4-input NAND
4068B	1	8-input NAND
4001B	4	2-input NOR
4081B	4	2-input AND
4071B	4	2-input OR
4070B	4	2-input EOR
4049UB	6	inverters
4050B	6	buffers
4016B	4	switch
4040B	1	12-bit counter
4060B	1	14-bit counter
4515B	1	4-bit decoder
4598B	1	8-bit latch
4536	1	programmable timer
4585	1	4-bit comparator

have overcome this problem), and the switching speed is dependent on the power supply voltage. For the SSI devices the maximum switching rate should be limited to 5 MHz with a 5 V supply and 10 MHz with a 10 V supply. Generally it is inconvenient to use anything other than a 5 V supply in a circuit which is to be connected to another system (such as a microcomputer), because of the necessity of ensuring that the levels used by the two systems are compatible. In spite of this, the advantages in terms of higher speed and greater output current capability may be sufficient to justify operation at 10 or 12 V, even if an extra circuit is required to change levels for interfacing to TTL.

The hi and lo logic levels of the C-MOS 4000B series depend on the power supply applied to the packages. The hi level is nominally equal to the positive supply voltage, and the lo level is nominally ground. To within about 1% these are the levels which the gates use for output, and the high input impedance of C-MOS gates means that these levels are also those that will exist on gate interconnections - so there is no level degradation produced by gates loading one another. However, the input of a C-MOS gate recognises the level change at almost precisely one half of the supply voltage; this is a considerable improvement over the TTL families and bestows upon the 4000B series a much better immunity to interference. Furthermore the switching time of a C-MOS gate is longer than the gate propagation time (ie. the time between input and output level changes) and this tends to result in C-MOS systems absorbing glitches and system noise - a particularly welcome feature after the horrors of some TTL circuits.

4.4 C-MOS and TTL together

For some reason one nearly always seems to need a TTL device in a circuit otherwise built exclusively with C-MOS devices - or vice versa. Fortunately interconnection is possible provided that the different impedances and current capabilities of the two classes are taken into account. If both devices are operating with a +5 V supply then C-MOS output will directly drive a single 74LS input or two 74L or 74ALS inputs (as shown in fig 4.6a). If standard TTL needs to be driven then one of the special TTL-driver C-MOS buffers must be used, as illustrated in fig 4.6b. An output from any of the TTL families can drive dozens of C-MOS inputs, although the TTL output must also be connected by a 2-10 k pull-up resistor to the +5 V supply line to ensure an adequate hi level value (fig 4.6c). The special TTL-drivers are required to interface TTL with C-MOS circuits that are operating at supply levels greater than 5 V, as in fig 4.6d. The 4050B (buffers) and 4049UB (inverters) are somewhat unusual in that they allow an input signal level to be greater than the supply voltage, although

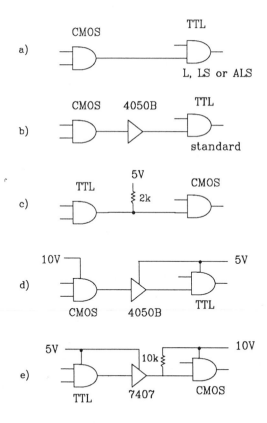

Fig 4.6 *Interfacing TTL and C-MOS devices. a) C-MOS (at 5 V) will drive the low power TTL families directly, but b) a high power buffer is required to drive standard TTL. c) TTL will drive lots of (5 V) C-MOS inputs with a pull-up resistor. When C-MOS is operated at 10 V, high voltage buffers must be used to interface with TTL, d) and e).*

the output hi level is determined by the supply voltage. The 7406 (inverters) and 7407 (buffers) have "open collector" outputs, requiring a resistance pathway to the positive supply voltage. This allows the hi output level to approximately equal that supply voltage, and these devices can be used to interface TTL with C-MOS operating at >5 V as illustrated in fig 4.6e.

The 74HCT C-MOS series deserves a special mention because of its remarkably high speed capability (50 MHz). Unlike the 4000, the 74HCT family is limited to a supply voltage of 5 V (7 V absolute maximum). However, it is capable of driving a standard TTL input or up to 10 74LS inputs, and may itself be driven directly by TTL levels without pull-up resistors (its logic input levels having been designed to be TTL compatible).

4.5 MSI circuits

Both the TTL and C-MOS systems have been developed well beyond the level of complexity required for the SSI gates introduced above. A wide range of ICs is now available using this medium scale integration (MSI) technology, and these make possible the manipulation of the 0 and 1 signal levels in almost any desirable manner. Discussing even a representative selection of the available MSI circuits is beyond the scope of this text, so that we shall make use of just two examples of this type of device. However, most manufacturers publish data books which provide details of all the ICs offered in their TTL and C-MOS ranges, and there are several books which discuss examples of the most popular devices in a thoroughly digestible manner (see bibliography).

4.5.1 A 12 bit counter

The first device that we shall examine has been chosen because of its particular relevance to laboratory applications of microcomputers, its function in that connection being discussed in chapter 7. The device is a binary ripple counter, a type of device available in a variety of forms both as TTL and C-MOS circuits. The particular form we shall discuss is the 12-stage binary ripple counter of the 4000B series, the 4040B. A schematic representation of the device is shown in fig 4.7. The device has an input (called the clock input, for reasons which will become apparent) which accepts C-MOS 0/1 levels, and twelve outputs (designated Q1 - Q12) which produce 0/1 levels. The connection called "reset input" is normally held at 0, ie. grounded. The levels of the twelve outputs can be regarded as representing a binary number of twelve BInary digiTS (bits). The pattern of twelve 0 and 1 levels (such as 001000101110, in which, by convention, the highest order bit - Q12 - is written first, just as for normal decimal numbers) forms a 12 bit binary number, which can have a decimal equivalent value in the range 0 (000000000000) to 4095 (111111111111). Thus ouput Q1 can be 0 or 1 and can contribute 0 (when Q1 is 0) or 1 (when Q1 is 1) to the value of the binary number. Output Q2 can contribute 0 (when Q2 is a 0) or 2 (when Q2 is a 1), output Q3 can contribute 0 or 4, and so on up to Q12 which can contribute 0 or 2048 (2^{11}). The number represented by the outputs (the "count") is incremented by one each time the clock input changes from a 1 to a 0 level. Thus the device counts the number of such "negative transitions" on the clock input, and as there is one negative transition each time the clock input goes through the cycle 0-1-0 (ie. each time a positive-going logic level "pulse" arrives at the input) the device can be used for counting pulses. The count can be reset to zero at any time by the application of a 1 level at the reset input, and counting starts again as soon as the reset input is returned to 0.

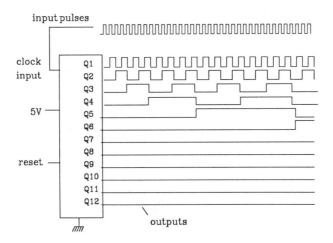

Fig 4.7   *The 4040B binary ripple counter, showing how the outputs (Q1 - Q12)*
change level in relation to a sequence of pulses applied to the input.

An alternative way of regarding the 4040B is as a divider, in fact it is often referred to as
a divide-by-4096 counter. The output Q1 changes only when the input changes from 1 to 0
(and not when the input changes from 0 to 1), so that when the input is "clocked", ie when a
continuous stream of alternating 0s and 1s is applied to the input as illustrated in fig 4.7, the
output Q1 changes half as frequently as the input - so it divides the input clock frequency by
2. The output Q2 divides the clock frequency by 2^2, ie. 4, and the output Qn divides the clock
frequency by 2^n. The highest order output is Q12, which divides the output by 4096 (2^{12}).

We can illustrate these two views of the 4040B by outlining two circuits which utilise
these principles. The first, shown in fig 4.8, is a fast timer which counts the pulses generated
by a 1.000 MHz crystal clock oscillator. Such oscillators produce highly accurate TTL-
compatible square waves, and are available for a wide range of frequencies from kHz to
MHz. The pulses are gated into the counter's clock input using an AND gate as a start/stop
switch, and the 12 bit binary output is displayed using 12 LEDs. (The LEDs are buffered
using the 12 4049UB high current inverters. Note that the inverters are used so that the LEDs
light on a lo level output. In this state the 4049UB can sink 5 mA when operating with 5 V
supplies. TTL buffers could be added if higher currents were required.) The Q12 output is
also connected to the clock input of a second 4040B, whose 12 outputs are displayed on a
second dozen LEDs. Thus the two counters form a 24 bit binary counting system which can
register the time between start and stop in the range 1 to 2^{23} (about 16 million) microseconds.
A simple manual reset switch is provided to reset both counters to zero. In this illustration the
start/stop switch is manually operated, although it would be more usefully activated by a
signal level derived from a laboratory apparatus, and the binary output is just a pattern of
lights, although it would be an ideal candidate for transmission to a microcomputer.
However, it is the 4040B's role which interests us at present - we shall return to these other
matters in due course.

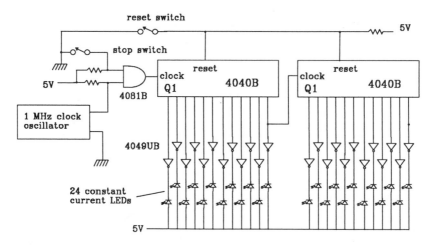

Fig 4.8 *A fast timer, using a 4040B to count the logic pulses from a 1.0 MHz crystal oscillator.*

It should be noted that the outputs of this circuit provide an example of a parallel digital signal - parallel because the 0/1 levels are carried on different conductors, side-by-side, but at the same time, and digital because the pattern of outputs is used to represent a number.

The second 4040B application example, illustrated in fig 4.9, is a pulse generator for producing 0-5 V oscillations at selectable frequencies of 6.4, 51.2, 409.6 and 3276.8 kHz. (The extension to a larger number of frequencies using the other outputs should be obvious.) In this example the signal from a 3.2768 MHz oscillator, which also provides the 4040B's input, and the output signals from Q3, Q7 and Q11 are fed to four 2-input NAND gates, the other input of each gate being set hi or lo by a "rate select" rotary switch. Only one of the NAND gates can have both inputs hi, and therefore only one NAND gate can have its output go lo. The outputs from the four NAND gates are combined using a 4-input NAND gate. This gate always has three inputs hi and one input changing at the selected frequency, so its output also changes at the selected frequency, being lo when all its inputs are hi and hi when one of its inputs is lo.

Circuits of this kind are valuable for providing pulses of highly reproducible duration, varying on/off times over a very wide range, or dividing down a pulse rate before it is measured on a ratemeter (while, incidentally, maintaining the statistical precision of the undivided pulse rate - see chapter 2 on shot noise).

4.5.2 The TRI-STATE[R] latch

The second device we shall discuss for handling digital signals is the TRI-STATE[R] latch, a device of particular importance for the transmission of digital signals to or from a microcomputer. While we shall discuss one particular device, the 74C373, the following points should be noted:

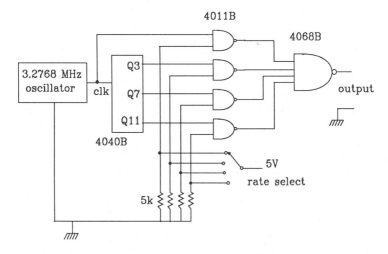

Fig 4.9 *Using a 4040B to divide a 3.2768 MHz pulse rate by a selected amount.*

1) Most of the principles of latching and TRI-STATE[R] output apply to a wide range of other devices designed to generate parallel digital outputs.

2) TTL versions of this device are available but are not suitable for some of the interfacing applications discussed in chapter 7.

Consider first an 8 bit parallel digital signal on 8 conductors, which we shall refer to as lines D1 - D8. Imagine that the signals on these lines are changing rapidly, as they would, for example, in the case of eight of the outputs from the timer circuit of fig 4.8. If we wished to examine the signals on the lines (eg. by looking at the display LEDs of fig 4.8) we would need to stop the signals from changing while the examination was in progress. Of course, stopping the timer of fig 4.8 also terminates our ability to continue with the time measurement, so that stopping the signals from changing by this method could be undesirable. An alternative technique is to pass the signals into a group of special buffers known as latches.

A single latch, illustrated in fig 4.10a, has an output, Q1, whose level follows the input, D1, as long as the level of the "not latch enable" pin is hi. When the level at the "not latch enable" pin is taken lo, then the output Q1 remains at the same level that it had when the "not latch enable" pin was hi. In other words the latch remembers the value of the input level at the moment "not latch enable" dropped from hi to lo. Now the words "not latch enable" tend to become a little cumbersome after a while, and the name latch enable is more commonly used to identify the pin. However, changing the name can cause confusion over whether latching occurs when the signal at that pin changes from a 0 to a 1 or vice versa. When the pin is called, correctly, "not latch enable", a hi or "true" level indicates that the latch is not enabled - ie the lock on the door between input and output has not been applied, so signals can still pass through. A lo or "false" level at that pin means that the latch has operated and data can no longer enter through the inputs.

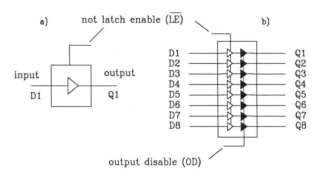

Fig 4.10 *a) A single latch. b) The 74HC373 octal latch with output disable facility.*

The 74HC373 is a C-MOS version of an octal latch and so has 8 latches in its 20 pin package, as illustrated in fig 4.10b. The 8 inputs (D1 - D8) feed 8 outputs (Q1 - Q8), and all 8 may be latched at the same moment by a hi to lo transition on the latch enable pin. Thus the value of an 8 bit number can be latched and examined on the outputs, Q1 - Q8, while the signals on D1 - D8 continue changing, allowing, for example, the timer of fig 4.8 to continue operating. When the latch enable pin returns to a hi level the outputs of the device will once again follow the inputs until another latch enable signal is applied.

Eight conductors can be squeezed into a fairly small space these days, but it would nevertheless be inconvenient to rely on a large number of multiwire cables for the transmission of parallel digital signals between items of laboratory instrumentation. Fortunately there is a remarkably simple alternative known as a "bus". A bus is a collection of conductors for parallel signals which is connected to several different sets of inputs and outputs, so that any of the input sets can receive signals from the bus, and any one of the output sets can transmit signals to the bus. For this system to work it is, of course, important that only one output at a time determines the value of the signal level on a single conductor. The signal levels we have met so far are not suitable for use in bus operation because there are only two allowed levels (hi and lo), and if one output on the bus is trying to force its attached conductor to lo, while a number of other outputs on the bus are trying to force the same conductor hi, then a lot of power is being expended with no certainty as to who will win.

One solution to this problem is to ensure that all outputs connected to the bus are open collector types. In this case each of the bus lines will be hi unless one of the device outputs is asserting a lo level on the line. Of course, it is up to the circuit designer to ensure that only one device is attempting to assert lo levels at any one time. The disadvantage is that C-MOS gates do not offer an open collector output variant, so that one of the TTL families must be used.

An alternative and elegant way around this difficulty, developed by National Semiconductor Corporation, is to introduce a third state in which an output can happily exist while some other output determines the level of the attached conductor. This third state is variously termed the "off" or the "high impedance" state. The important point is that in this third state the output does not effect the signal level on the conductor attached to it. Although this technique is called TRI-STATE[R] logic, it is important to remember that the signal levels remain the same as conventional TTL or C-MOS levels (depending on the device); it is a third output state which has been added (an off state) and not a third signal level.

In order that the outputs may be directed into the high impedance state, TRI-STATE[R] devices have a pin connection called "output disable". With a lo level applied to the output disable pin, the devices' outputs behave just like any other logic outputs. When a hi level is applied to this pin all the devices' outputs are switched into the high impedance state. Of course, if this facility is not required at all then the output disable pin can be permanently grounded. The 74C373 has TRI-STATE[R] outputs and these are compatible with C-MOS and low-power TTL inputs (with a fan-out for the latter of 1). Similar devices are available in most of the other 74 families, both TTL and high-speed C-MOS, although naturally the inter-family compatibility will vary (see section 4.4).

Figure 4.11 shows three octal latches with their inputs derived from a suitable 24 bit parallel digital signal source, such as the outputs of fig 4.8. The outputs of the latches are connected into a common bus which also connects to eight buffered LEDs. The latch enables are connected together and operated by a single switch which can be manually operated to latch the data on the 24 input lines. The three output disables are connected to separate "examine" push switches, so that whichever switch is operated the corresponding 8 bits of latched data are placed on the bus and displayed on the LEDs. It is a good idea to study this

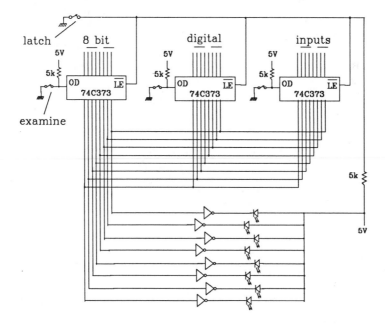

Fig 4.11 *Connection of the outputs of several 8-bit TRI-STATE[R] devices to a common bus - in this case feeding 8 buffered LEDs through high current inverters.*

circuit until its operation is fully understood. There are more of these to come as they form the basis of multifunction computer interfaces.

4.6 Generating logic levels

We have already met the TTL compatible crystal oscillator packages suitable for generating 0/5 V pulses. These and 5 V dc supplies provide the main source of logic signal levels. However, a major requirement in the application of digital and logic circuits is often the generation of logic level signals from some other kind of signal. For example, many types of transducers generate pulses which are counted when the transducer is used for measurement purposes. However, the pulses generated by the transducer are rarely precise changes from 0 to 1 levels and back, so that some kind of interface circuit is normally required to translate the non-logic signal levels to the required logic values. Several types of device are available for this interfacing role, but one of the most widely used is the comparator.

4.6.1 The comparator

A comparator is superficially similar to an operational amplifier - in fact op-amps can be used as comparators in some cases. However, a comparator is designed to produce an output logic level dependent on the differential input at its inverting and non-inverting inputs. (Generally the output is TTL compatible, and should also serve C-MOS operating with 5 V supplies if helped with a pull-up resistor.) An example is shown in fig 4.12, where the transfer function (ie. a curve showing the output as a function of the differential input) illustrates that the output is a logic level 0 whenever the non-inverting input is more than a millivolt lower in potential than the inverting input, and a logic level 1 whenever the non-inverting input is at the higher potential. As the comparator's input signal is a differential signal, a comparator can be used for generating logic level signals in response to the variation of an analog input signal compared with some preset voltage level. A simple example of this use is shown in fig 4.13, where an analog input signal, V_{in}, is applied to the non-inverting input and is compared with the "threshold" level applied to the inverting input. Note that the

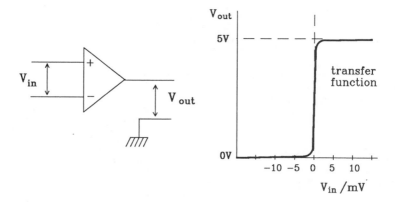

Fig 4.12 *Circuit symbol and transfer function of a voltage comparator.*

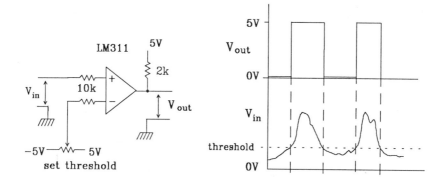

Fig 4.13 *The use of a comparator as a threshold detector. The circuit produces a logic 1 output while V_{in} is greater than the threshold level.*

311 has an open collector output, in this case taken through a 2 k resistor to a TTL hi level, so providing an output compatible with both TTL and 5 V C-MOS. By changing the hi reference and resistor value (to, say, 10 V and 10 k) this device could also drive C-MOS operating with 10 V supplies. Not all comparators provide open collector outputs.

Comparators suffer from many of the imperfections discussed for operational amplifiers, and generally speaking require relatively high input bias currents. However, comparators are designed to be used without feedback loops (although some feedback can be useful to introduce an element of hysteresis into the transfer function), and to produce an output which can change level fairly rapidly through the range 0 to 5 V (most comparators cannot generate a negative output voltage, even though many require a negative supply voltage). The imperfection which is peculiar to comparators is a variation in the rate at which the output can change as a function of the differential input voltage - a variation which is often different for positive and negative input steps. This imperfection is characterised by manufacturers in diagrams showing the response time (which is actually the time delay between an input change and an output change) for various input overdrives (the amount by which the differential input voltage exceeds zero). A typical selection of such curves is shown in fig 4.14, illustrating the kind of data which needs to be considered before a particular comparator

Table 4.5 Characteristics of some typical voltage comparators

Device no	306	311	319	339	360
Devices per package	1	1	2	4	1
bias current /μA	25	0.25	1	0.25	15
response time /ns					
(100 mV overdrive)	40	200	80	1300	16
TTL fanout	10	5	2	1	2
C-MOS > 5 V?	no	yes	yes	yes	no

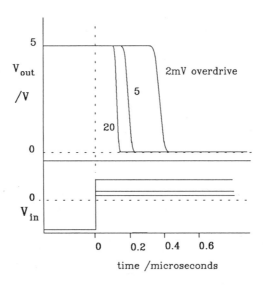

Fig 4.14 *The variation of the response time of a typical voltage comparator for various input overdrives.*

is chosen for a circuit. Clearly a slow device with the characteristics showns in fig 4.14 would not be a good choice for use in a circuit which would need to produce pulses at a rate close to 10 MHz. The basic properties of a number of comparators are collected in table 4.5.

Figure 4.15 shows a circuit for a light detector designed to produce a logic pulse when the photomultiplier senses a scintillation (a brief flash of, in this case, about twenty photons). The PM tube is being operated with a (virtually) grounded anode, and the pulses of negative charge arriving at the inverting input of the transresistance amplifier produce positive-going voltage pulses at its ouput, A. These signals are passed through an input resistor to the inverting input of a fast comparator, whose non-inverting input is held at a suitable threshold voltage by VR1. Under dark conditions the voltage at the comparator's inverting input is lower than the threshold voltage, so the comparator's output, B, is at a TTL 1 level. When a scintillation produces a voltage pulse at A which exceeds the threshold voltage, the comparator's output changes to a 0 and remains there until the voltage at A has decayed to a value below the threshold. Thus the signal at B consists of negative-going pulses (ie. the level goes from 1 to 0 when a pulse is initiated; the voltage level never goes negative). The width of each pulse depends on how long the voltage at A remained above threshold, so the pulse widths are not constant, although this does not matter if the pulses are only to be counted. Indeed measurement of the pulse widths may be used to determine the intensities of detected scintillations.

The analog signal at A is useful if the signal needs to be examined on an oscilloscope, or if pulse height analysis is to be performed. When connections of this kind are to be made to circuits handling ac or pulse signals it is important to check that the connection does not affect the operation of the rest of the circuit or, worse, grossly distort the analog output, by the introduction of additional capacitive impedance. With the component values shown the

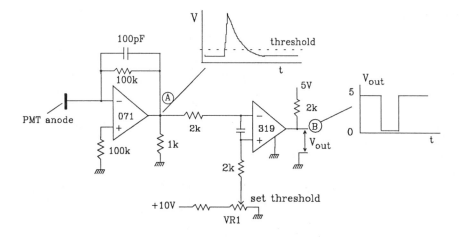

Fig 4.15 *The production of logic pulses from a photomultiplier tube monitoring scintillations. A transresistance amplifier is connected to a comparator with a threshold level set by VR1. The threshold may be set to determine the minimum scintillation intensity which gives rise to an output pulse.*

circuit counts scintillations at up to about 10^4 per second with dead-time losses of <3%. Faster devices could be used to improve this figure, although to achieve low dead times at pulse rates above 1 MHz requires the use of narrow (<100 ns) pulses with even shorter (<20 ns) rise times. Great care needs to be taken with the layout of circuits intended to handle pulses in this frequency region.

A considerable improvement in speed over the circuit of fig 4.15 can be obtained by the use of a single IC device designed specifically for the conversion of low-level charge pulses to logic pulses. A circuit using one such device is shown in fig 4.16, where the device is the Amp-Tek PAD (pulse amplifier discriminator) A111. The circuit does the same job as that described above - except that it counts up to 10^5 Hz without significant dead time. The device is excellent, and expensive, and very delicate - touching its pins with a test signal can destroy it.

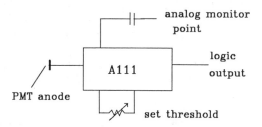

Fig 4.16 *The simple arrangement possible when using a charge amplifier discriminator, such as the AmpTek A111, for the detection of photomultiplier pulses.*

4.6.2 The monostable, astable and bistable

Another valuable device for the production of logic pulses is the monostable, a device which prefers to retain its output at one level although it can be forced into the other level for a preset period of time. Most of these devices are useable in two slightly different ways, depending on connections made to the device, and the devices are referred to using the broader name "multivibrator." Firstly, they can be used to produced a single logic pulse of specified width in response to some input trigger. An example is shown in fig 4.17a, where a 4047B (C-MOS) multivibrator is being used as a monostable, producing output pulses in response to positive-going transitions at the +trigger input. (A -trigger input is also provided so that the device can respond to negative-going transitions.) Both positive going and negative-going output pulses are available from the device's complementary outputs, and the width of the output pulse is 2.48 CR s. The monostable finds application in reducing a stream of variable width pulses to a series with uniform width, and in producing logic pulses from non-logic type signals - although this particular device is triggered by "edges" of voltage change, which have to be quite fast (typically <10 μs).

The second way in which this device may be used is as an astable multivibrator and in this guise it produces a continuous stream of pulses. The 4047B is capable of astable operation only with a 50% duty cycle (ie. the times spent by the output in the hi and lo levels are equal), so the output appears as a square wave. An example of this mode of operation is given by fig 4.17b, where the frequency of the output pulse stream is governed by CR and the period is equal to 4.4 CR s. The 4047 can be used to provide pulse streams in the frequency range 1 kHz to 1 MHz and is therefore a useful device for the production of clocking signals, although the frequency stability is typically only a few percent - which is several orders of magnitude poorer than can be achieved with a crystal oscillator.

A closely related device is the bistable or flip-flop circuit. There are in fact several varieties of these, although they share the properties that their outputs can remain hi or lo (ie, they have two stable output states), and that their outputs can change levels only at particular times in relation to a clock signal applied to the device. It is this latter property which defines the flip-flop as the basic circuit of a system known as clocked- or sequential logic, and the fundamental building block of the bulk of MSI circuits, such as the binary ripple counter we have already discussed.

A typical flip-flop is illustrated in fig 4.17c. This one has two inputs in addition to its clock signal input, and these are called the J and K inputs (although there does not appear to be any particular reason for this nomenclature) and the device a JK flip-flop. The important properties of a JK flip-flop are:

1. If the outputs are to change they can only do so when the clock signal
   goes from lo to hi.

2. The outputs are determined by the J and K inputs according to the rules
   summarised in the truth table in fig 4.17.

In addition JK flip-flops such as the C-MOS 4027B have two other inputs which force

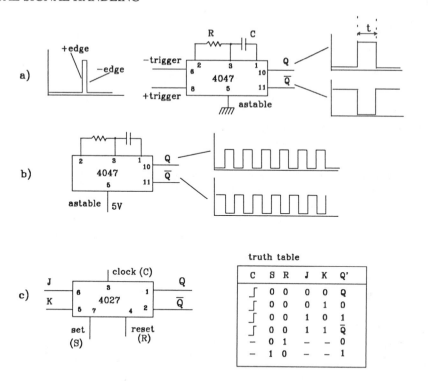

Fig 4.17 a) Monostable operation of the 4047B, producing a single pulse each time the device is triggered. b) Astable operation, generating a continuous stream of pulses as a square wave. c) The 4027B bistable JK flip-flop device.

the outputs into particular states irrespective of the clocking signal. Thus a 1 on the SET input forces output Q to a 1 (and the complementary output to 0), and a 1 on the RESET input forces Q to a 0.

This type of device has a wide variety of applications in addition to forming the basis of more complex devices. For a start it is the basic "divide-by-2" unit, in that with J and K both hi, the output changes once for every two level changes (ie. hi-lo-hi cycle) on the clock input. (We will use this property in fig 7.17) Furthermore the device can be used to synchronise one level change with another, a particularly valuable feature in timing applications. For example, if J=0, K=0 and Q=0, then a change to J=1 will not result in a change to Q=1 until the clock signal changes from lo to hi. Similarly if J=0, K=1 and Q is forced hi by a brief SET=1, then the output, Q, returns to 0 only when the clock input changes to a 1. (We will use this property in fig. 7.16.)

4.7 Analog/digital interconversion

Digital techniques offer ways of handling and transmitting signals rapidly and reliably, and are highly immune to the effects of interference. Microcomputers are exclusively digital, but the laboratory world remains heavily dependent on analog signals. Most transducers generate analog outputs, and many "results" are presented in an analog form - such as spectra, chromatograms and other chart records - even when quantitative data is digitally recorded. Clearly the ability to interconvert analog and digital signals is of major importance in the laboratory applications of microcomputers.

Performing these interconversions electronically requires electronic circuits which map a dc signal (usually a voltage, say, 0 - 10 V) onto an n-bit binary number between 0 and 2^n-1. This mapping is illustrated in idealised form and on a highly enlarged scale in fig 4.18a, where the stepped function indicates the actual mapping. The broken line drawn between 0 V, binary 0, and 10 V, binary 6, represents the mapping which would result if fractional digital values were allowed. The points at which the mapping is exact are those at which the lines cross. At all other points the mapping is in error, by up to the equivalent of half a digital unit, ie. $\pm 1/2$ LSB (a least significant bit, 2^0). For an 8 bit binary number (0-255) this error is about 0.2% of the maximum signal value, and for a 0-10 V analog range is equivalent to ± 20 mV on the analog scale. In general the error is equivalent to 1 part in 2^n for an n-bit conversion, and is called the quantisation error of the conversion. Its effects are most serious for small signals; for example, with a 10 V full scale range an analog signal mapping to the 8 bit digital value 2 may be 78 ± 20 mV, an error of 25%.

This kind of conversion error is unavoidable when converting between a continuous analog signal and a discrete digital signal. Its effects may be reduced by operating in the range where the digital signal is a substantial fraction of the 2n-1 maximum value, or increasing the resolution of the conversion (ie. the number of bits used for the digital signal).

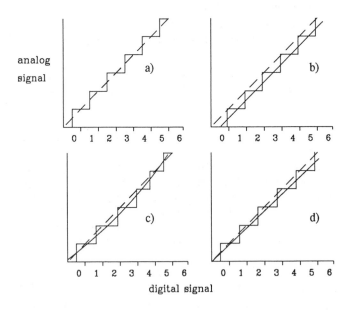

Fig 4.18 *The mapping between an analog and a three bit digital signal, showing a) the inevitable quantisation error, b) a zero error, c) a nonlinearity error, and d) a full scale error.*

In practice the conversion error is more serious than suggested by fig 4.18a because of imperfections in the circuits which bring about the conversion. These additional sources of error are many and complex, and a full description of them is beyond the scope of the present text. However, the effects of the three most important contributions to the additional conversion error, illustrated in fig 4.18b-d, are described briefly below - to sow a few seeds of healthy scepticism about the accuracy of the conversion devices considered later.

Figure 4.18b shows the effects of a zero error, where the actual mapping is displaced from the ideal mapping. This error arises from input offset effects and may be easily corrected by offsetting the analog signal. In fig 4.18c we see the effect of a non-linear mapping function which can arise through circuit imperfections. At the point of maximum deviation from ideal mapping (which is -1/2 LSB in this example, the difference between the broken line and the solid curve) the converted signal may be in error by up to 1 LSB, although at the point of contact between the broken line and the stepped function the mapping is, of course, exact. So a specified non-linearity error of +/- 1/2 LSB implies a possible conversion error of ± 1 LSB. Figure 4.18d illustrates the effect of a -1/2 LSB linear error (also called a full scale error) in the slope of the mapping function. A full scale analog signal correctly maps to a digital signal of 6, but slightly smaller signals may have errors of up to 1 LSB.

4.7.1 Digital to analog conversion

There are several techniques for implementing digital to analog conversion. The principle of one of the simplest is shown in fig 4.19. The circuit shown is for a 4 bit digital to analog converter (DAC) accepting 0 or 5 V signals at its digital inputs, b0 - b3 (corresponding to binary inputs of 0 - 15), and producing an analog output voltage in the range 0 - 9.125 V. The circuit consists of a summing amplifier producing an analog output given by

$$V_{out} = b3 + b2 / 2 + b1 / 4 + b0 / 8 \text{ V}$$

There are two major difficulties with this approach. Firstly it requires a number of resistors with very precise resistance ratios - as these determine the linearity of the conversion. This becomes difficult to achieve as the number of bits increases, and in practice alternative techniques are used in the manufacture of single package DACs. Secondly this technique relies on the actual voltage levels at the hi digital inputs for the value of the analog output. This would not be acceptable as, for example, the TTL logic levels can vary over quite substantial ranges, and in practice the digital input hi levels need to be converted to a precisely defined analog reference voltage level.

The most popular single package DACs are those using a technique known as the R-2R ladder (a variant of the fig 4.19 circuit). Most devices working on this principle use an 8 or 10 bit digital input. An alternative type of DAC is the monolithic DAC, which utilises bipolar transistors to generate scaled currents from the digital signal levels, and a transresistance amplifier to produce an output voltage proportional to the sum of these currents. Monolithic DACs are available as 8, 10 or 12 bit devices.

More sophisticated techniques allow up to 16 bit (or even 18 bit) digital data to be

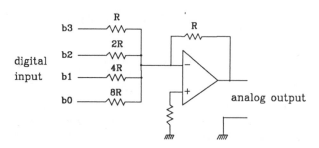

Fig 4.19 *A typical digital to analog converter circuit.*

converted, although at a significant cost. Some DACs produce an output voltage which is proportional to the product of an input reference voltage (or current) and the digital signal, with the reference voltage variable over a wide range. Such "multiplying" DACs allow the analog output range to be varied or scaled, and this can be extremely useful where wide range analog output is required with only limited resolution.

As DACs tend to use analog output circuits there are limitations on the rate at which the analog output voltage may change. Furthermore the higher resolution DACs may require a significant time to carry out the conversion process and produce a stable output. Generally conversion is initiated on the application of a logic level to a "convert" pin, and the time required after this for the analog output to become stable (ie. not to change by more than the equivalent of 1/2 LSB) is called the settling time. The characteristics of some typical DACs are collected in table 4.7.

4.7.2 Analog to digital conversion

One of the many techniques used for converting an analog input signal into a digital output signal is known as parallel conversion (or flash conversion). This technique is used in some of the fastest analog to digital converters (ADC) and an example of a converter based on parallel conversion is shown in fig 4.20. The circuit illustrated allows an analog input

Table 4.7   Characteristics of some popular DACs

Device no	type	resolution /bits	settling time /μs
ZN435E	R-2R	8	0.5
ZN428E	R-2R	8	1.25
DAC0800	multiplying	8	0.1
MC3410F	multiplying	10	0.25
AD7520	multiplying	10	0.5
AD7542	multiplying	12	2
AD7534	multiplying	14	0.8
DAC703JP		16	2.5

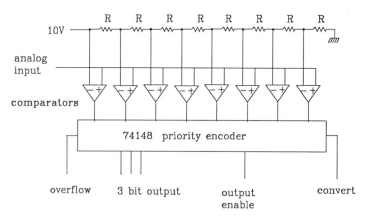

Fig 4.20 *A typical analog to digital converter circuit based on the parallel conversion technique.*

signal in the range 0 - 10 V to be converted into a three bit binary output in the range 0-7, when the input enable (convert) line is taken lo. The chain of equal value resistors provides eight equally spaced reference voltages between 1.2 and 10 V inclusive, and the eight comparators produce outputs of 0 or 1, depending on whether the voltage at their inputs is greater than or less than the corresponding reference voltage. Thus an input signal of 5 V would produce 0s from comparators 1-4 and 1s from comparators 5-8. This pattern of 0s and 1s is then encoded into a three bit binary number by a "priority encoder", which produces a binary output determined by the highest of the eight inputs to be at the 0 level (irrespective of the signals on the lower number inputs). The outputs may be set equal to 7 at any time by taking the output enable line hi (note that the 74148 does not have TRI-STATE outputs).

Parallel conversion ADCs are available with up to 8 bit binary outputs and are ideal when very fast conversions are required. Unfortunately they are expensive. The more humbly priced ADCs are slower devices and generally require an "initiate conversion" pulse on one pin, producing a "busy" signal on a second pin until the conversion is completed. One of the commonest ADC types uses a technique known as successive approximation. The analog input voltage is compared with an analog signal generated by a DAC from a digital value which the device adjusts, by successive approximation, until the two analog signals are within 1/2 LSB. The digital value arrived at now forms the ADC output. This technique requires that the DAC changes its output value n times (for an n-bit converter), and so needs at least n DAC setting times to achieve conversion. To avoid the risk of the analog input signal changing during this time interval, successive approximation ADCs usually sample and hold the analog input signal at the start of the conversion cycle. Thus noise on the analog input (especially short duration spikes) can produce fluctuations on the digital output. Successive approximation ADCs are generally highly regarded for accuracy, although some can suffer from a problem known as "missing codes" - which is, essentially, producing the wrong answer once in a while.

The other main approaches to analog to digital conversion are based on the charging of a capacitor. This may involve using standardised current pulses being counted into a capacitor

Table 4.8   Characteristics of some popular ADCs

Device no	type	resolution /bits	conversion time /μs
CA3300	parallel conv.	6	60 ns
TL507C	single slope	7	1000
ZN439E	succes. approx.	8	5
ZN448E	succes. approx.	8	10
8703CJ	charge balance	8	1250
TDC1007J	parallel conv.	8	33 ns
AD7581	8 channel sa	8	20
AD573J	succes. approx.	10	30
AD574J	succes. approx.	12	25
ICL7109	charge balance	12	33 ms

until the voltage across the capacitor exceeds the analog input voltage (single slope integration). The count then forms the digital output. Alternatively a charge balancing technique may be used. The most popular variant of this involves using a current proportional to the analog voltage to charge the capacitor for a fixed time, followed by the discharging of the capacitor with a constant current and digitising the time required to empty the capacitor (dual slope integration). The problem with single slope integration is that the conversion accuracy is limited by the accuracy and stability of the capacitor and the comparator used. On the other hand it does provide one of the best techniques for producing a uniform spacing between adjacent portions of the conversion mapping. Dual slope integration results in very accurate conversions which are not particularly sensitive to the value or stability of the charge storage capacitor (the capacitor is usually "on chip"). The charge balancing techniques are all fairly inexpensive, but slower than the successive approximation method. Some typical ADCs are listed in table 4.8.

4.8 Serial digital signals

Although we will not consider a specific serial data system until chapter 8, it is appropriate to consider the basics of serial digital signals here. The parallel digital signals we have met so far have consisted of signal levels on several different conductors - each one representing a bit of the digital signal. As all the bits of a parallel signal are available at the same time, time is not an important element in determining the byte value of an 8 bit parallel signal. Serial signals on the other hand consist of a series of signal levels applied sequentially to a single conductor, so that it becomes essential to agree on the time span that each bit of the signal occupies. A typical serial signal representing an 8 bit value is illustrated in fig 4.21, where the duration of each bit is indicated.

Communicating byte values between devices in the form of serial signals is attractive because it can be achieved with a simple twin cable (signal conductor and ground). However, it does require agreement between the devices at either end of the cable on when each

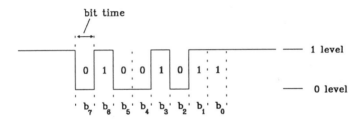

Fig 4.21 *One example of serial digital data. In this case the data represents an 8 bit number.*

transmission starts and on the duration of each bit. We will return to these aspects of serial transmission in considering the RS232 data communication standard in chapter 8. For now we confine our attention to the generation of serial signals and serial/parallel interconversion.

While in principle any sequence of level changes on a single conductor may be regarded as a serial digital signal, for our purposes the important serial signals are those used to represent 8 bit digital values. The simplest way of producing a serial signal of this type involves generating it from an 8 bit parallel signal, using an IC device known as a shift register. A typical device, illustrated in fig 4.22, is the 74165, an 8 bit PISO (Parallel Input/ Serial Output) shift register. It has 8 parallel inputs (identified as A - H) through which an 8 bit parallel digital signal may be latched into the register by a hi-to-lo transition on the shift/ load pin.

When the shift/load pin goes lo the serial output, Q_H, goes to the level stored in the register's H bit (and the complementary output is at the other level). When shift/load goes hi the register is able to shift its bits one bit to the right on the rising edge of each of the pulses applied to its clock input. The old content of H is discarded and H receives the old G bit, G receives the old F, F receives the old E and so on. A receives whatever level is present on the serial input pin, so if this is permanently grounded A becomes 0. As the output Q_H produces the new value of H, this output shifts through the sequence of levels representing the bits H down to A during 8 clock cycles. In our example the serial input pin is grounded, so once the 8 bits of data have been shifted out of the register Q_H remains at 0 (and the complementary output at 1).

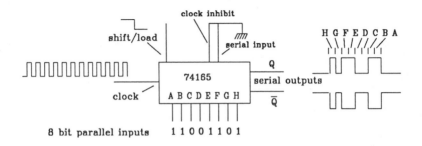

Fig 4.22 *The 74165 8 bit parallel input/serial output shift register.*

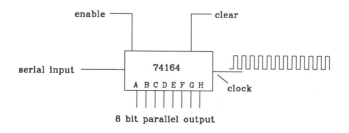

Fig 4.23 *The 74164 8 bit serial input/parallel output shift register.*

The reverse translation process is achieved using a Serial Input/Parallel Output (SIPO) shift register, such as the 74164 illustrated in fig 4.23. This device has 8 outputs which are determined by the corresponding bit values in its register. A lo signal applied to the clear pin sets all bits to 0. The device has two serial inputs. When both inputs are hi then a 1 is placed into the register's A bit on the rising edge of the next clock pulse when the register shifts right and the old content of A is moved to B, the old content of B is moved to C and so on. The old H is discarded. When either of the two serial inputs is lo a 0 is forced into the A bit on the rising edge of the clock signal. This arrangement makes it possible to use one of the serial inputs to enable the other (as if either one is held lo the 8 parallel outputs become 0s after 8 clock cycles). Of course, the desired 8 bit parallel output only has its 8 bits all present during the eighth clock cycle, so it is up to the user to ensure that this byte is transferred to a latch at the right time.

4.10 Some useful logic controlled circuits

There are a number of devices which are particularly useful for the control aspects of instrumentation and which can be activated by logic levels or pulses. One of these is the logic level power MOSFET, which is a development of the field effect transistor and allows a substantial current through the transistor to be turned on or off by the application of a logic level to its gate. A typical application of a power MOSFET is shown in fig 4.24a, where a logic level is used to open or close a solenoid valve. In this example a lo input logic level turns the MOSFET (and so the valve) off, while a hi input turns it on. We have used a number of these circuits to operate valves in flow analysis instruments, in which the valves open when 12 V is applied across the solenoid, drawing a current of around 200 mA. In other instruments substantially larger voltages and currents may be handled by suitably rated MOSFETs. For example, the RFP10N15L can, when mounted on a suitable heat sink, handle a current of up to 10 A and withstand a voltage across its terminals of 150 V. A MOSFET can turn on or off much more rapidly than a mechanical switch or relay (typically in a microsecond or less), and with a high level of reproducibility. However, when currents flowing through inductive loads (such as solenoids) are switched off suddenly a large back voltage may appear briefly across the load, so it is generally wise to protect the switching device by placing a diode across the load to short out this back voltage as shown in fig 4.24a.

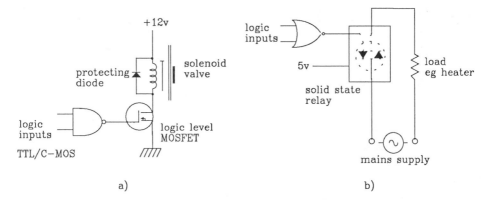

Fig 4.24 *Two useful switches controlled by logic levels: a) the power MOSFET, and b) the solid state relay for mains power switching.*

A related device is the solid state relay used for switching on or off alternating currents. These are available in forms capable of switching up to 40 A (again when mounted on a suitable heat sink) and are useful for operating mains powered devices such as lamps and heaters, as illustrated in fig 4.24b. It is important to bear in mind that solid state relays rely on the fact that the ac waveform of the passing curent reaches zero (twice per cycle) to enable them to turn off. Consequently they cannot be used to switch direct currents. Another consequence of their operation is that they turn on and off not at the instant that the controlling logic level is applied to their input, but when the switched waveform next passes through zero. Thus, when switching a 50 (or 60) Hz mains supply, there may be a delay of anything from 0 to 25 (or 16) milliseconds before the power is removed. Generally delays of this order are not important in controlling ac powered devices, but it is as well to be aware of their existence.

A valuable circuit element for instrumentation which invloves mechanical movement is the stepper motor controller IC. A stepper motor is an electric motor which has four windings, and when these are energised in the correct sequence the motor's spindle rotates by a single step of a well defined angle (typically 1.8° or 7.5°). While they are not particularly fast motors (in terms of revolutions per minute), they do have the advantage of being able to move at a precisely controlled rate and a precisely controlled number of revolutions - hence their widespread use in graphics plotters, monochromators and robot arms. It is perfectly possible to construct a circuit to handle the correct winding energising sequence, but a stepper motor controller can more easily provide the sequence each time a specified logic transition is applied to its input. Furthermore, some controller ICs allow aspects of the sequence to be specified by logic levels applied to other inputs, as illustrated in fig 4.25. Thus the step direction (ie clockwise or anti-clockwise) and the step size (actually a whole step or half a step) may usually be specified. While some stepper motors can be operated from a 5 V supply, most require higher voltages (typically 12 V) and in any case all but the smallest motors draw more current than can be handled with dil-type ICs. Consequently the controller must usually be followed by power transistors, MOSFET switches or a driver IC capable of handling the current required by the motor.

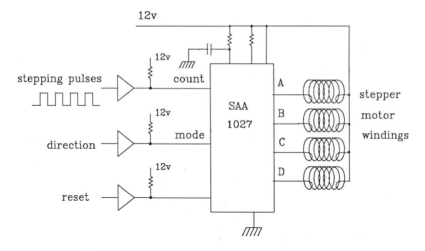

Fig 4.25 *A stepper motor controller which steps the motor 1 step for each stepping*
pulse. Note that the controlling logic signals are 0 or 5 V levels, while the
SAA 1027 uses 0/12 V signals.

As we shall be meeting more complete circuits in later chapters, the example shown in
fig 4.25 is for a relatively low powered stepper motor driven directly from a controller IC (an
SAA1027) which is itself operated from a 12 V supply. This controller can drive motor
windings which require up to 0.5 A, although we only used it for 100 mA motors
connected to a monochromator grating. The logic levels (0/5 V) provided as inputs to this
circuit are used to drive open collector buffers (EQ gates) which may have their outputs
pulled to a voltage higher than 5 V. Thus the controller IC receives its logic inputs as 0 or 12
V levels. A logic level specifies the direction of each step, while the motor steps each
time a lo-to-hi transition arrives at the "count" input. The reset input is used to ensure that the
controller can be reset to a combination of winding currents usually specified as the idle state
of the motor. (Even when at rest a stepper motor is still drawing current. If this current is
turned off then the motor's spindle is free to rotate.)

The circuit in fig 4.25 is quite capable of receiving and processing input pulses far more
rapidly than any stepper motor is able to respond to the resulting current impulses.
Furthermore the rate at which a motor can respond depends very much on how fast it is
moving when the current impulses arrive. Thus when the spindle is at rest its own inertia
results in a low maximum stepping rate. The "pull-in" rate is the maximum stepping rate at
which the motor can start from rest without losing steps. A typical low power motor may be
able to pull-in at around 100 steps per second, while some of the more powerful models can
manage 1000 s^{-1}. The consequence of driving a stepper motor at too high a stepping speed is
that it misses steps, so that its spindle either doesn't rotate as fast as the controller imagines
it should or it doesn't rotate as far (ie as many steps) as the controller wishes. In either
case the important fact is that the person or computer operating the controller loses track of
the motor's condition, and the main advantage of using a stepper motor is lost.

CHAPTER 5

THE MODERN MICROCOMPUTER

Digital signal handling circuits continued to develop from the MSI devices described in chapter 4 until it became possible to produce thousands of gates in a single device - a technology known as large scale integration (LSI). During the late 1970s a number of companies began using these devices to produce small desk-top machines which had many of the capabilities of the available computers: the microcomputer age had dawned. Since that time a wide variety of microcomputers has become available. Some are designed for the business market, for use in word-processing, record keeping and accounting. Others are aimed primarily at the personal or hobby market - and often that means game playing and some useful software for graphic displays. With the exception of some of the Hewlett Packard range and a number of specialist machines, few are actually designed to be used for instrumental purposes in scientific laboratories, although many can fulfill this role surprisingly well.

5.1 Bits and bytes

Most of the early microcomputers used in laboratory applications were 8-bit computers, i.e. the numbers stored in the computer's memory and used for communication with external instrumentation are handled as "bytes" of 8 bits (BInary digiTS) each. Bytes are of fundamental importance in understanding many aspects of all modern computers - even those which handle number in groups of 16 or 32 bits - so a short discussion of their nature and use follows.

Table 5.1   The 8-bit byte and its equivalent decimal value

bit number:	b7	b6	b5	b4	b3	b2	b1	b0
values	1/0	1/0	1/0	1/0	1/0	1/0	1/0	1/0
contribution	2^7	2^6	2^5	2^4	2^3	2^2	2^1	2^0
ie.	128	64	32	16	8	4	2	1
	decimal range 0 - 255							

Table 5.2   Examples of 8-bit binary numbers

b7 b6 b5 b4 b3 b2 b1 b0	b7 b6 b5 b4 b3 b2 b1 b0
0 0 0 0 0 1 1 1	1 1 1 1 1 1 1 1
= 4 + 2 + 1	=128+64+32+16+8+4+2+1
= 7	=255

A byte representing a binary number is shown in table 5.1. The byte is stored within the computer as eight digital signals, each of which can be either a 1, taken to represent a binary 1, or a 0 representing a binary 0. When shown diagrammatically as in table 5.1 the eight bits are referred to as bit 0 - bit 7, or b0 - b7. The binary value of the byte is then just a list of eight 0's and 1's, and the equivalent decimal value can be calculated by summing the contributions of each bit, remembering that a 0 bit contributes decimal 0 while a 1 bit contributes 2 raised to the power of the bit number. Thus a 1 for b0 contributes decimal 1 (2^0), a 1 for b2 contributes decimal 4 (2^2) and a 1 for b7 contributes decimal 128 (2^7). The range of values which can be held as a single byte is therefore 0-255. Some examples of 8-bit bytes are shown in table 5.2.

Computers are able to use bytes in a number of ways: a) Bytes can be treated as numbers and the computer can perform arithmetic with them. b) Two different bytes can be compared to see if they have the same value, or whether one has a greater value than the other. c) Bytes can be stored within the computer or read in from, or transmitted out to, an external device, such as a printer, a magnetic disk unit or a laboratory instrument. It is important to realise that a particular 8-bit binary code may also serve a variety of different purposes, representing a value, a pattern, a character (letter, number or symbol), a binary coded decimal (BCD - see

Table 5.3   Typical 8-bit codes

dec	BCD	hex	char	binary
65	41	41	a	01000001
66	42	42	b	01000010
90	—[a]	5a	z	01011010
50	32	32	2	00110010
44	—	2c	,	00101100
13	—	0d	<cr>[b]	00001101
10	—	0a	<lf>[c]	00001010

Notes: a) not a valid BCD code
          b) carriage return character
          c) line feed character

appendix 2) pair of digits, a hexadecimal (hex - see appendix 3) pair of digits, and so on. Some examples of the quantities represented by byte values are given in table 5.3 and more detailed equivalence tables are provided in appendices 1-3. Finally d) bytes are used as the coded instructions (called machine code) which control the operation of the computer, so that a computer program is actually a string of bytes treated in a particular way by the computer.

Fortunately there is no necessity to plunge into the kind of programming mentioned in d), as the microcomputers likely to be of most value in the laboratory may be equipped with commercial software which allows the user to program the computer using a high level language such as BASIC. Furthermore high level languages enable the user to treat numbers within the computer as decimal quantities, unrestricted by the small range of values which can actually be stored in a single byte. However, it will be necessary for us to understand something of the structure of a computer and the manipulation of bytes if we are to make use of the computer's facility to transfer bytes between itself and a laboratory instrument, and we shall return to this subject in section 5.2.

High resolution graphics and several aspects of the byte handling systems we shall discuss below make use of the rapid and simple setting or testing of particular bits of a byte. For example, parallel interface port connectors, through which bytes may be transferred into and out from the computer, generally have at least one control line output in addition to the eight data lines, and it is necessary to be able to switch the level of this line between hi and lo to operate a peripheral connected to the port. In many cases the control line is actually connected so that it resembles one bit (or more) of a specific byte associated with the port's circuit, so that from the programming point of view we need to be able to change the value of one bit of that byte while leaving the other seven bits unchanged.

Testing or changing specific bits of a byte may be accomplished on most micros in

Table 5.4 The AND operation and its implementation in BASIC

The bytewise effect		BASIC examples
X	15	IF (X AND 1)>0 THEN 100
		goes to 100 if b0=1
0 0 0 0 1 1 1 1		
		Z = X AND 64
Y	170	IF Z = 0 THEN GOTO 2000
		goes to 2000 if b6=0
1 0 1 0 1 0 1 0		
		L = X AND 15
X AND Y 10		H = X AND 240
		L & H are nybbles of X
0 0 0 0 1 0 1 0		
		IF X >= 128 THEN 1000
		goes to 100 if b7=1
(15 AND 170) results in 10		

Table 5.5   The OR operation

The bytewise effect	BASIC examples
X 15	
——————	Y = X OR 128
0 0 0 0 1 1 1 1	set b7 of Y to 1
Y 170	
——————	IF (A OR B)=1 THEN 100
1 0 1 0 1 0 1 0	go to 100 if b0 of A
X OR Y 175	or b0 of B = 1
——————	Y = X OR 32
1 0 1 0 1 1 1 1	set b4 of Y to 1,
	rest of Y as X
15 OR 170 results in 175	

BASIC with the aid of Boolean operations such as AND and OR. Such operations involve comparisons between corresponding bits of two bytes and produce a result which is stored in a third byte. For the AND operation, illustrated in table 5.4, the result of the BASIC instruction Z = X AND Y is found by setting each bit of Z to a 1 if the corresponding bits of both X and Y are 1s. Those bits of Z for which the corresponding bits of X and Y are not both 1s are set equal to 0s. The AND operation is most frequently used to pick out particular bits of one byte by ANDing with a second byte in which the bits of interest are 1s. Thus to discover whether bit 0 of X is a 1 we could take X AND 1 (since decimal 1 is 00000001) and the result would be zero if b0 of X was a 0, but non-zero if b0 of X was a 1. Some samples of BASIC coding for testing or extracting bits of X are included in table 5.4. The AND operation is also useful for setting particular bits of a byte to 0 without affecting other bits of the byte. For example X = X AND 127 sets b7 of X to a 0 without affecting the value of the other seven bits (127 is 01111111 in binary).

The OR operation is illustrated in table 5.5, where the result of the BASIC instruction Z = X OR Y is found by setting each bit of Z to a 1 if either (or both, it's an inclusive OR operation) of the corresponding bits in X or Y are 1s. Those bits of Z for which the corresponding bits of both X and Y are 0s are set equal to 0. The OR operation is most commonly used to force particular bits of a byte to be 1s without affecting the values of other bits in the byte. For example, X=X OR 128 sets b7 of X to a 1 without changing the value of any of the other seven bits (128 is 10000000 in binary). Some versions of BASIC (including Microsoft's BASICA) also provide an XOR operator which allows an eXclusive OR comparison between two bytes (ie. the bitwise result is a 0 if both bits are 1s, *cf.* section 4.1). Unfortunately the logical operators provided on a number of older micros do not support bitwise operations and so cannot be used in the manner described above.

5.2 Microcomputer elements

The elements of a typical microcomputer are shown schematically in fig 5.1. The system consists of a microprocessor unit (MPU) which controls the flow of bytes around the system and actually performs any calculations, comparisons and byte transfers required by its program, and some memory ICs (more LSI devices) in which the bytes are stored for use by the MPU. The MPU in early microcomputers was generally a 40-pin IC, of which the 6502 (produced by MOS Technology), the 6800 (Motorola), the 8080 (Intel) and the Z80 (Zilog) were the most numerous examples. The actual MPU used in a micro is not of great importance unless a good deal of byte level or low-level assembler programming is envisaged. It is true that some MPUs can operate faster than others, but for many purposes a microcomputer's behaviour under the control of a high level language will also be dominated by the quality of the user's program and the efficiency of its BASIC - not just by the clocking frequency of its MPU.

Microcomputers contain at least two types of memories. First there's read-only memory (ROM), which contains a permanently stored program provided to control the way in which the computer operates when switched on - and in some cases the program which accepts the user's instructions in BASIC. Then there's read/write or random access memory (RAM), which is used to store the user's program and data numbers, generally for only as long as the power is switched on (although micros with non-volatile memories, which retain their RAM contents even after the main power is switched off, are becoming more common). A typical microcomputer may have, say, 20000 bytes (20k bytes) of ROM containing BASIC among other things, and 32-640k bytes of RAM (although a "k" (kilobyte) is actually 1024 (2^{10}) in this context). A program cannot alter the values of bytes stored in ROM, but only read the values that are already there, while variables and program bytes stored in RAM can be changed at will.

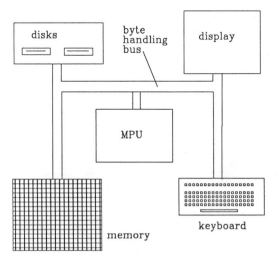

Fig 5.1 *Schematic representation of the elements of a typical microcomputer.*

In early microcomputers the type of RAM used was known as static RAM. This retained its byte contents until the power was turned off, but tended to consume a large amount of electrical power and occupied too much space. Later systems have tended to use dynamic RAM, which retains byte values for only a few milliseconds and so needs to be refreshed every couple of milliseconds by the application of an electrical signal to each bit. (The popularity of the Z80 MPU for early hobby micros was due in part to the fact that it provides these refresh signals, thereby reducing the amount of additional circuitry required.) Dynamic RAM became more attractive as it required less power per byte, and allowed more bytes to be stored on a given chip area. Recent advances in chip technology have resulted in improved static RAM becoming available, and a number of recent micros (such as the Compaq 386) contain a megabyte or so of this new static RAM. However, from the micro user's point of view, there is very little difference between RAM types. Several other types of memories are available for bulk storage, and for laboratory use a magnetic disk (see below) is probably the most useful for the longer term storage of programs and data. Over the next few years we may expect to see other "add-on" memories, such as large non-volatile RAM, optical disks and bubble memories, become less expensive and more readily available.

Most microcomputers have a keyboard which is the primary pathway of communication (input) to the computer from the user, allowing the user to type in a program (in BASIC for example) which is then stored in the RAM until the computer is instructed to obey the program - again with a command from the keyboard. Most also have some kind of alphanumeric display device (like a calculator display) or a video output for connection to a television or a video monitor. Some early micros, like the PET, and many of the semi-portables had a built in video monitor, while later micros (such as the Compaq, Toshiba and Panasonic portables) have liquid crystal, electro-luminescent or gas plasma discharge display screens. The display device is the primary pathway of communication (output) from the computer to the user, for displaying results and error messages and allowing the user to check what he types in on the keyboard.

Of fundamental importance in understanding the operation of the computer is the structure of the byte handling pathways or "busses" within the machine. Figure 5.2 shows the

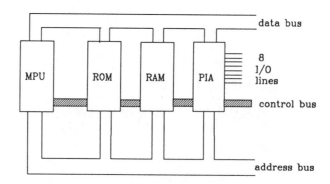

Fig 5.2 *The principal byte handling and control pathways and major components of a typical microcomputer.*

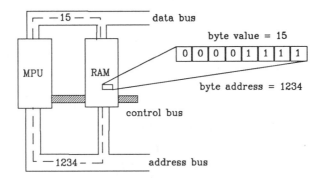

Fig 5.3 *Referring to a specific byte within the micro's memory from BASIC. In most BASICs the function PEEK(1234) returns the value of the byte at address 1234, while the instruction POKE 1234,15 stores the value 15 into the byte.*

way in which the MPU, ROM, RAM and byte input/output system are inter-connected. The MPU is actually connected to three busses; one is the data bus along which bytes are transferred to or from the MPU, and another is the control bus which the MPU uses to control other parts of the microcomputer (for example, to specify whether it wishes to read or write a byte to or from memory). The duration of the digital signals is controlled by a clock signal derived from an oscillator and running at a frequency typically in the range 1 - 20 MHz. One of the signals generally available on the control bus is a clock signal of some kind, although usually a modified form of the MPU's clock input signal. The third bus is the address bus, on which multi-bit binary numbers are planted by the MPU. An address is used primarily to specify which byte of ROM or RAM is being referred to by the MPU. Where a 16-bit address bus is used addresses can range from 0 to 2^{16} -1 (65535 otherwise known as 64k), so that the MPU can address up to 65535 bytes, while a 24-bit address bus allows an MPU to address more than 16 Mbytes (megabytes; 1 Mbyte = 2^{20} = 1048576 bytes).

When programming in BASIC the user does not have to worry about which bytes are being used to store data. However, a particular byte of memory can be referred to in a BASIC program, as illustrated in fig 5.3, usually (in Microsoft BASIC) by using the instructions Y=PEEK(X), which collects the number (data) stored in the byte addressed as X and stores it into Y, and POKE X,Y, which stores the data value held in Y into the byte with the address X. (Of course, POKing ROM doesn't work.) The version of BASIC supplied with some large memory size computers (such as the IBM PC) may limit the range of addresses accessible through PEEKs and POKEs (typically to a 64k range). The BBC microcomputer's version of BASIC uses the query indirection operator (?) for the equivalent of both PEEKing and POKing, viz. Y=?X and ?X=Y.

5.3 The video display

An important component of any microcomputer is the system provided to enable the computer to produce an output display. Virtually all the desk-top microcomputers currently available are designed to produce their display as a raster scan on a video monitor or TV, as

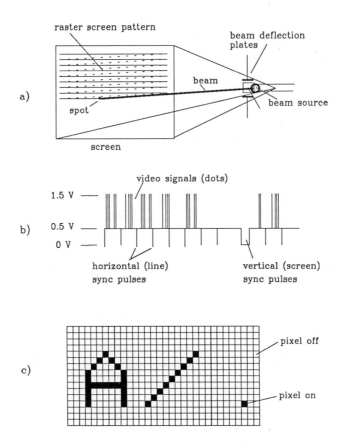

Fig 5.4 *(a) The elements of the raster scan video display system. (b) Typical video signals. (c) The formation of a typical character from pixels.*

illustrated in fig 5.4a. (There is an alternative approach, known as vector imaging, but this has remained confined to the more expensive end of the small computer market.) The exact method adopted for generating the display varies rather widely and is unlikely to have a direct impact on the laboratory use of a micro. Consequently we shall discuss the topic in a general way and ignore the details of the relationship of the video output system to the rest of the byte handling circuitry. Standard video signals consist of three signals: a vertical synchronisation (sync) pulse, the end of which signals the start of scan from the top of screen; a horizontal sync pulse, which signals the start of scan from the left of each scan line; and the intensity level signal, which determines the beam intensity - and so the brightness of the spot on the screen. The intensity signal is a voltage level between 0.5 V (for black) and 1.5 - 2 V (for white - or maximum brightness of the screen's phosphor colour), while the sync pulses are drops from the 0.5 V level to 0 V. When all three signals are combined together the signal is called composite video, an example of which is shown in fig 5.4b, and this is suitable for direct connection to a video monitor. When the video signal is used to modulate an ac signal of the appropriate (UHF) frequency the resultant signal may be used to drive a normal television receiver.

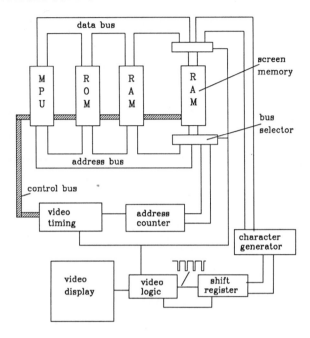

Fig 5.5 *Outline of the character position mapping technique in which a byte from screen memory is decoded by the video circuitry to produce a series of pulses to generate the required dot pattern on screen.*

A raster scan produces images on the screen by varying the intensity of the scanning electron beam, and in most microcomputer applications the intensity variation allowed is only on or off. The on/off condition is determined by the values of a succession of logic pulses, representing 0s or 1s but scaled to the video signal levels, with each hi level pulse producing a single dot on the screen. For the output of a character the video signals are formed in such a way that the pattern of dots will appear on the video screen at a precisely determined time after the start of the electron beam scan, and therefore at a precisely determined posistion on the screen.

There are three broad approaches to the generation of video signals for a screen display. The first is character position mapping, illustrated for a monochrome system by the block diagram illustration in fig 5.5. The central feature is a block of RAM (screen memory) into which the computer's operating system places bytes representing the characters which will be displayed on the video screen in response to PRINT instructions etc. This block of RAM is also being frequently examined (typically 50/60 times a second) by a second hardware system generally referred to as the video logic system. This system extracts from the screen RAM each byte in turn (at times when the RAM is not being used by the MPU), and passes it to a pulse generating circuit often made up of a character generator ROM connected to a shift register. This circuit translates the byte code into a serial pattern of pulses which are then converted into video signals, forming a pattern of dots on the video screen.

The screen display is generally arranged as a rectangular matrix, varying from 16 lines of

40 character positions (for TV display systems) to 25 lines of 80 character positions (for the more expensive video monitor systems). As at least one byte of RAM is required for each character position, the "screen memory" may occupy anything from a few hundred to 2k bytes of RAM. The addresses used for screen memory may be found in the microcomputer's memory map, usually specified in the machine's reference manual. If the micro's operating systems allow direct access to the screen memory from BASIC, then the contents of character positions may be examined or changed using PEEK and POKE instructions repectively. Otherwise the screen display is produced by simple PRINT or LIST statements.

Each screen character position has its dot pattern determined by the contents of one byte (in simple monochrome systems) of screen memory. Each byte can hold values in the range 0 - 255, although less than half of these are required as codes for all the upper and lower case letters, numbers and punctuation marks. As a result most micros use the "spare" values to provide dot patterns which are useful for generating graphic displays on the screen. For example, many home computers provide for a range of graphics characters, such as lines, squares, bars and in some cases the card suite symbols. These graphics characters allow the user to build up simple but crude pictures on the screen to represent graphs and histograms, etc. The detail which can be included in such pictures is small because of the limitation in the number of character positions.

A second character based display system has developed from the ROM generator technique described above. Many micros (eg. the Commodore 64, Amiga and Macintosh) allow bit patterns for character generation to be stored in RAM, so that the dot patterns defined for byte values in screen memory may be changed by programming. The approach adds considerably to the range of shapes which may be produced on the display screen, although retaining the concept of character positions within the display. Some modern micros allow both the character generator and the area of RAM used as screen memory to occupy any part of the available RAM.

A third display scheme, known as the bit mapped display, uses a bit in RAM to represent every individual dot position on a screen. This approach is invariably adopted for graphics displays, and requires only that the program creating the graphics image is able to set individual bits in RAM to 0 or 1. When characters are generated using this technique the bits which contribute to a single character are no longer confined to a single byte (or even to adjacent bytes), so that the screen display cannot be interrogated simply by PEEKing screen memory. On the other hand a virtually unlimited number of patterns may be generated and positioned anywhere on screen. This allows multiple fonts, scientific symbols and graphics diagrams to share the same screen display. The cost of this flexibility is memory usage. For a monochrome graphics display of 320 (horizontal) * 200 (vertical) dot positions, 64,000 bits are required, occupying 8k bytes of RAM, while a 640 * 400 display would require 32k of RAM.

Many popular micros now produce signals for multicolour display on colour televisions or monitors, and these originate from three different intensity signals, red, green and blue. High quality monitors for colour displays are called RGB monitors and have separate inputs for the three colours' intensities and one for the sync signals. Mixing the colour levels in the proportions of 1:2:4 (for blue, green and red respectively) to generate colour video levels, and adding in the sync pulses, produces the colour equivalent of the composite video signal. This may be used to drive the single input colour monitors associated with CCTV and audio-visual systems. The increased demand for high quality colour displays on microcomputer systems has led to the widespread adoption of a technique involving the direct connection of logic signals carrying red, green, blue and intensity data (RGBI) from the computer to the appropriate monitor. This approach is used for the IBM PC and many of its clones, and is usually identifiable by the 9 way D socket on the computer for connection to the monitor. It is important to bear in mind that both RGB and RGBI monitors are in current use and incompatible, and that different PC display adapters use different beam scan rates (and so require different or switchable-rate monitors).

Of course, colour displays require more RAM for the screen memory than monochrome displays, as information about the colour of characters or pixels must be stored. The techniques for handling the RAM arrangements for colour images are broadly similar to those adopted for monochrome displays, although the variety of implementations has become sufficiently large that discussion here must be limited to a small number of examples included in our discussion of the PC display adaptors in the next chapter.

High resolution images on video monitors can be photographed fairly easily, although the quality of the image on a standard TV screen is rarely adequate. For some popular microcomputers the video signal used to generate the display may be fed to a device called a video copier, which at the touch of a button (or command from the computer) can produce a paper photocopy of the pixel pattern on the screen - thus allowing the rapid production of hard copy (although, generally, of lower quality than could be obtained from a laser printer or slower pen plotter). In some cases this technique is able to produce colour images both directly on paper or film, or by rapidly transferring them to video recorders.

5.4 Disks

A very valuable component of most modern micros is the floppy disk unit. A floppy disk is a magnetic storage medium on which programs and data can be stored for long periods (ie. years), and yet copied back into the computer in a few seconds. Floppy disks come in several sizes (approximately 3.5, 5.25 and 8 inch diameter), and data can be stored on them in a variety of ways - many computer manufacturers naturally choosing a different storage density and layout to ensure complete incompatability with any competitor's product. Floppy disks holding 100k - 1M are now commonplace, although it is likely that floppies with much greater capacities (eg 10M) will become increasing popular. The drives which hold the floppy disks and arrange the recording and playback of information are normally built into the micro's cabinet, and may record on one or on both sides of the disk. The disk itself is positioned in the drive when required, and removed and stored in a dustproof sleeve when not in use.

While modern floppy disks are remarkably reliable they may eventually fail - through scratching, loss of magnetic coating, accidental erasure or physical accident - and the data on them may be irretrievably lost. (In the author's laboratory a pack of ten disks was in frequent use for a period of two years, during which time not a single failure occurred. Then, within a period of two weeks, three of the disks failed.) Thus a golden rule for the disk user is "always keep a backup copy of anything you would hate to lose". This means copying the contents of a disk onto a spare disk. If data is being added to continually this may require daily copying of a disk's contents - although strictly speaking it is only the latest data which should need copying to a backup disk - and for really valuable data a second, perhaps less frequent, backup copy may also be advisable. The difficulty is that copying a disk takes time, and the larger the amount of information on the disk the longer the time required for making backup copies. Small capacity disks (eg. 100 kbyte) may take a couple of minutes to backup on a computer fitted with two disk drives, while a 1 Megabyte disk may take 10 minutes on a single drive machine. It is not so much that the time is a problem, but that some people tend to forget to make long backups more readily than they forget to make short backups.

Hard disks (sealed units in which the disk spins much more rapidly than in the floppy system), also known as Winchesters, have become increasingly popular and are available at reasonable prices (ie a few hundred dollars). These can hold much larger quantities of data than most floppies (eg 10 - 100 Megabytes), so keeping backups becomes more of an art, although their principal attraction is that loading and saving programs and data may be accomplished much more rapidly than with floppies. High capacity hard disks are especially useful if several micros can be connected to the same disk unit, or if a particular laboratory application requires the storage and recall of really massive amounts of data (eg. pattern matching of spectra or chromatograms). Cassette and cartridge tape units are available for backing up hard disks on most popular micros.

5.5 Peripherals

Nearly all of the major microcomputer manufacturers supply a range of peripheral devices which may be attached to their micros. Printers are particularly valuable for the production of a "hard copy" of results, although they are also extremely useful when one is working on program development - along with a massive stock of paper to allow for many listings of the program before all the bugs are ironed out. Printers are also available from a number of independent manufacturers, and often these are cheaper or of better quality than otherwise similar printers supplied by the micro's manufacturer. Daisy wheel printers function like typewriters, physically imprinting specific characters onto paper one after the other. Dot matrix printers produce their characters in the form of patterns of dots created by a column of (usually) 5, 7 or 9 "wires" striking the ribbon, spraying powdered ink-patterns, or developing heat sensitive or electrosensitive paper. While the print quality of dot matrix printers is not generally as high as that from the daisy wheel printers (although the later 24-wire matrix printers are capable of quite acceptable print quality), dot matrix printers are probably more useful for laboratory applications. They tend to be faster than daisy wheels, and many are capable of producing "high resolution" dot graphics (albeit rather slowly), allowing hard copy of graphs, histograms, spectra etc. to be recorded.

Laser printers (based on photocopier technology) are becoming popular as their prices become more reasonable. Most of these print a dot matrix, although typically based on a resolution of 300 dots per inch (dpi), so one needs a magnifier to distinguish the printing from "proper" printing. Many laser printers allow the character patterns to be provided by the computer ("downloading" the "font") or by a plug-in cartridge, so that a wide variety of type styles and sizes may be used. Some laser printers have graphics printing capability (also based on up to 300 dpi), although it is important to remember that a full page of graphics requires a very large number of bits, and so a large internal memory for the printer. Most laser printers have limited memories and cannot print a full page of graphics at 300 dpi.

It is also important to bear in mind that the various printing fonts and graphics printing capabilities of the dot matrix and laser printers rely heavily on control bytes being sent from the computer to the printer. Much commercial software is sold with "printer drivers" which provide the necessary control codes for a specified number of printers (typically including several Epson models and Hewlett-Packard Laserjet printers). However, not all printers are supported by all programs, and most libraries associated with language compilers cater only for ASCII character output. Persuading a quality printer to do anything other than print characters (in the default font with default margin settings) using a high level language like BASIC or C, can turn into a substantial programming effort. Sales images of high quality graphics appearing from easily prepared BASIC programs should be treated with considerable caution.

In the last couple of years there has been a dramatic growth in the number of graphics plotters available for connection to micros. One or two rather well known companies have produced such instruments for many years, but now there is a range of reasonably priced alternatives (eg. $500 - 1000 from companies such as Epson, Roland and Watanabe). These devices enable one to program the computer to draw virtually anything, and to add printed characters. Essentially the program gives the plotter two sets of x and y coordinates and the plotter draws a line between them. The resolution of available plotters (ie. the smallest movement the pen can make) varies from about 0.01 to 0.5 mm, and plotters are available for paper sizes from A4 (approximately 8*12 inches) up to A2 (approximately 16*24 inches). Some plotters are quite intelligent and can be instructed to draw smooth curves through a number of data points, draw and annotate axes of graphs, or change pen colours during operation. Others rely on computer software for plotting under the control of a BASIC program. In this case the necessary routines are generally supplied by the plotter's manufacturer, and may not be usable with compiled BASIC.

Some of the more expensive plotters will allow the user to position the pen manually and then transfer the x and y coordinates of the pen into the computer. This technique is useful for "digitizing" data from a graphical form (eg. old spectra, chromatograms or diagrams), and digitising attachments for some low cost graphics plotters are available from some companies. For most micros digitizing "tablets" (such as the Summagraphics range) are available to offer this function without the plotting facility, the user tracing over the required diagram with a special pen or puck while the diagram is held flat on the tablet's surface. The ultimate in digitising is surely the page scanner, which is a bit like a laser printer in reverse, in that it can provide a bit image of whole pages with a resolution of up to 300 dpi. The only two requirements for using these scanners (such as those produced by Canon and Hewlett Packard) are a computer with a substantial memory, and some good software which can utilise the bit patterns produced. Some scanners are provided with software which is able to read pages of text and produce a file of character codes, which may then be incorporated into word processor files. Unfortunately few scanners are currently provided with software which is able to treat diagrams other than as massive bit image patterns.

5.6 The micro families

In the late 1970s most of the available microcomputers were based on one of three microprocessors, the 6502, 8080 or Z80. These processors all use 8-bit internal registers for processing the bytes and are called 8-bit processors; so the computers based on them - the Apple, BBC, Commodore 64, PET, Spectrum and so on - are referred to as 8-bit microcomputers. These 8-bit processors also share another important characteristic - a 16-bit address bus, providing an addressable memory space of 64k.

While many of the early 8-bit micros were useful machines, the principal problem they faced was that they were all different; they used different processors, different disk formats, different display systems and different programming techniques for fundamental tasks such as checking the keyboard. A consequence of this and the limited memory space (which resulted in many programs being written in machine code which could not easily be moved from one type of micro to another) was that commercial programs were either poor in quality or expensive - because the market for any one program was relatively limited. This in turn led to relatively small markets for the computers themselves.

In 1981 IBM entered the microcomputer market by introducing the Personal Computer (PC), a machine which could have 256k of memory. The microcomputer market boomed as commercial organisations bought these respectable machines, and software houses quickly produced programs for the PC. The PC was based on the Intel 8088 processor, an MPU which uses 16-bit internal registers, so the PC was the first major 16-bit microcomputer. The 8088 still used an 8-bit data bus and was therefore only able to move bytes around one at a time, but it did have a 20 bit address bus and so could theoretically address up to a megabyte of memory. Unfortunately the 8088 has only 16-bit internal registers, so it is not easy to generate a 20 bit address and with hindsight we can see that this deficiency has held back even more rapid advances in software. However, even the limited 256k of memory in the PC was an improvement over the majority of earlier machines and the result was a significant improvement in the quality of microcomputer software.

The original PC was designed as a modular micro. The main circuit board held the processor, up to 256k of RAM, variable amounts of ROM (some of which could contain BASIC), electronics for controlling the keyboard, and very little else. Refinements, such as electronics for running a display screen, disk drive or printer had to be obtained as extra circuit cards which could be plugged into expansion slots on the main board. This (then) unusual way of selling microcomputers contributed significantly to the success of the PC, because it forced program writers to develop techniques for producing programs which would run on PCs having a variety of configurations (ie. things plugged into the slots).

If IBM's name helped to ensure sales of the PC, so its excellent Technical Reference Manual ensured that many other companies soon followed with copies and improved copies of the computer - generally referred to as PC clones - which ran the same software and used the same additional circuit cards. This in turn led to other companies producing improved add-on circuit cards, such as the Hercules display boards which provided a much better display quality for both text and graphics than the original products.

Several variants on the PC have appeared during the last five years, but all are based on members of the 8086 family of processors (consisting of the 8088, 8086, 80186, 80286 and 80386), and maintain a very high degree of software compatibility. The enormous success of the PC and clones (over ten million have been sold) has ensured that, whatever the shortcomings of the processor and other features of the design, the PC-type microcomputer will be around for a number of years to come - in spite of the second generation PC (which has different slots) being released by IBM in 1987. So large a share of the microcomputer market has been taken by PCs that the bulk of our discussion of interfacing will be illustrated with reference to the PC, and a more detailed examination of the PC is included in the next chapter.

Two important microprocessors utilise 32-bit registers, so microcomputers constructed around them presumably qualify as 32-bit micros. Several such machines are based on the powerful Motorola 68000 family of processors. The 68000 processors have a 16-bit data bus (so bytes can be moved around two at a time) and a 24-bit address bus (allowing 16 megabytes of memory to be addressed). The processors also carry out important operations significantly faster than the 8086, so 68000-based micros have a number of advantages over the PC. Apple have produced several versions of their machine, the Macintosh, and now offer models with 512k, 1M or 2M of RAM and floppy or hard disks. The Macintosh had the distinction of being the first major 68000 system and early versions had a very high quality monochrome display which made it particularly attractive for office use. However, its potential for laboratory use was severely limited by its restricted interfacing facilities. The later Macintosh II has a full colour display and greatly improved facilities for interfacing to external devices.

Commodore produce the Amiga micros, which incorporate very impressive high speed colour graphics facilities and a powerful operating system (AmigaDOS). The first Amigas had only 256k of RAM (although expansion to 512k was simply a matter of plugging in an expansion pack) and limited disk facilities. However, in addition to the usual printer and serial ports, the Amiga does have an 8-bit parallel interface port which retains a high degree of compatibility with earlier Commodore computers. The new versions of the machine offer powerful facilities, hard disks, IBM PC compatible add-on cards and considerable scope in laboratory applications.

The Atari ST and its derivatives have a Macintosh type high quality monochrome display or not-such-high-quality colour, large RAM sizes, a wide range of input / output ports, and a low price tag. It makes an excellent computer for personal use, although in most of its varieties it is less convenient for laboratory use, largely because of its single unit construction, large footprint and external hard disks.

These 68000-based machines all have some significant advantages over the PC. Unfortunately they share one major disadvantage - they haven't sold in such vast numbers. While their advantages could have eventually led to the downfall of the PC, the advent of the 32-bit 80386 processor from Intel may have tipped the scales back in favour of the PC. The 80386 has a 32-bit data bus (allowing bytes to be moved four at a time) and a 24-bit address bus (giving a 16M address space). Furthermore, the processor is capable of running the vast amount of 8088, 8086 and 80286 software written for the existing 16-bit PCs. Several companies now offer PC-compatible machines based on the 80386 (such as the Compaq 386), machines offering 2M RAM, fast, high-capacity hard disks, and quite respectable graphics displays. Some of these micros run programs four times faster than 80286-based machines and sixteen times faster than the original 8088-based PC. Furthermore, the PC compatibility extends to the expansion slots, which allows the wide range of circuit and interface cards developed for the PC to be used - a particularly attractive feature for the laboratory user.

Ultimately all of the above machines will be supplanted by others which offer much greater speed. Micros based on reduced instruction set processors (RISC machines such as the Archimedes) and processors which may be operated in parallel (such as the Transputer) offer rather dramatic enhancements in virtually every aspect of operation. It is likely that communication with laboratory equipment will be less dramatically affected, because the speeds of measurement and control signals are generally governed by requirements beyond the control of micro manufacturers. However, it will undoubtedly be great fun to have super fast micros displaying their abilities.

5.7 The programming language

The majority of currently available microcomputers may be programmed in the high level language BASIC, which generally can be learned in a few hours by almost anybody. Most micros can use an alternative language, and some (principally the PCs) have a wide range of languages available, including Algol, APL, Assembler, C, COBOL, Comal, Forth, FORTRAN, Lisp, PASCAL or Prolog. Each language has both advantages and disadvantages, and supporters and critics. It is probably true to say that most languages are designed for application to particular classes of problems, so most will offer the user a compromise between ease of use, efficiency and suitability for any other application. However, it is not our objective to survey programming languages, nor to provide a lesson in any of them. Because of the overwhelming preponderance of BASIC as the prime language of most available low cost micros, our discussion of the application of microcomputers to laboratory instrumentation will be carried out largely in terms of software written in BASIC, and it will be assumed that the reader is familiar with a version of this language. However, the systems discussed in the remainder of this book could have been prepared in any one of a variety of languages (some were implemented in C), so although BASIC is used for the purposes of discussion, it is not the author's intention to endorse the choice of this particular language.

Unfortunately it is necessary to emphasise immediately that there are many different versions of BASIC, some which use a different syntax for a similar instruction, and others which for a given syntax produce a different result or effect. One of the most widely used forms of BASIC is that written by Microsoft, although even this has become bifurcated to the extent that a program written in Microsoft BASIC on one micro cannot be guaranteed to work on another micro equipped with Microsoft BASIC. Thus the Commodore 64, Apple II, TRS80 and IBM-PC all use BASICs written by Microsoft, although there are differences between the versions - and a variety of versions have been produced for the PC clones by different companies. Hewlett Packard produces many different computers, but their BASICs are different again, as are those of the Amiga and the Acorn-produced BBC microcomputer.

In some cases the differences between dialects of BASIC are small (eg. GOTO and GO TO may not both work, PRINT USING and PRINT AT commands may not be available, multiple assignments - eg. A=B=C=0 - may not be permitted, functions may not require brackets, and the operator used for exponentiation varies considerably), but in others they may be more fundamental, requiring a considerable amount of reprogramming time if a translation from one to another is required. For example, A$(3) may mean the third character of string A$, or, more commonly, element 3 of a string array A$(I), and, while most BASICs use the functions LEFT$, MID$, etc. for string handling, others use an X$(1 TO 5) syntax. Some BASICs have MAT functions available for handling matrices, and some allow REPEAT....UNTIL, WHILE.....END and/or IF...THEN.. ....ELSE structures. Rather more confusing, while most micros use either GET or INKEY$ to test for a key pressing on the keyboard, and return a value whether a key has been pressed or not, the GET and GET$ functions on the BBC microcomputer and the GET instruction on the Apple all halt the program and wait for a key to be pressed before continuing. Most micros handle logical arithmetic in the normal bitwise manner, but the Apple II and old TRS80 both use their own special forms of logic and do not permit simple bit manipulation in BASIC.

Most of the popular micros can be programmed at the byte level using the machine code of the MPU. (There are differences between the machine codes for different MPUs, but once you are familiar with one it's not difficult to get used to any other.) While such programming can be carried out by POKing the numerical codes directly into RAM using a BASIC loader, this is a tedious business for all but the smallest blocks of code. It's generally much easier to write programs using nmemonic codes known as assembler language, as this also permits the use of names (rather than RAM byte addresses) for variables and parts of the program. Once an assembler language program has been written it can be translated into the equivalent byte values using a program known as an assembler. It generally takes longer to write, test and modify an assembler language program than an equivalent program written in BASIC. However, assembler language programs should run many times faster than BASIC ones - unless, of course, the speed is limited by an external factor, such as data transfers to a peripheral or a laboratory instrument.

For laboratory applications it is often possible to have the best of both worlds, writing assembler routines for small, but often used, parts of a program (such as the routines for collecting bytes from a laboratory instrument), while retaining the ease and flexibility of BASIC for those parts of the system which perform calculations and print or display results. Mastering the elements of an assembler language is actually not as difficult as it may at first appear. The situation is eased by the fact that, in many cases, only a handful of the assembler language instructions need to be used. Some micros (eg. the BBC micros) allow assembler language instructions to be written directly into BASIC programs, while others require that assembler language routines are loaded separately or the appropriate codes POKEd into RAM. Most allow the use of machine code subroutines, where a BASIC program transfers control to a machine code routine using a CALL or SYS instruction (eg. SYS 32070), with its parameter pointing to the address of the machine code routine, and many allow additional parameters in the form of BASIC variables to be passed between BASIC and the machine code. The machine code routine instructions are then carried out and control returned to the BASIC program by an RET (or RTS) instruction, the assembler language equivalent of BASIC's RETURN.

Between the worlds of BASIC and assembler language lies another pathway which may lead to speed without pain. A compiler is a program which translates a high-level language program into a machine code program. In the days of punched cards, paper tape, FORTRAN and Algol, all programs ran on "real" computers where compiled, and it was the compiled, machine code form of the program which actually carried out the calculations and produced the results. BASIC changed all that. A BASIC program sits in the modern microcomputer's RAM and is interpreted BASIC instruction by BASIC instruction as it runs. Thus in a FOR loop, the BASIC instructions in the loop are interpreted each time the program cycles through the loop. Compiling a BASIC program converts it once and for all into a machine code form, so there is no interpretation while the program is running. In principle then the compiled program should run much faster than its BASIC form, although not as fast as a program written in assembler in the first place.

This author's experience of compilers for some low-cost microcomputers has been disappointing, particularly when their performance is judged against the advertising claims. In some cases the speed increase over normal BASIC has been small (eg. 2. One magazine review of a BASIC compiler noted that some programs tested actually ran slower when compiled!). For commercial use one of the attractions of a compiler is that it produces relatively tamper-proof code, and speed may be a less important aspect of the compiler's operation. In many cases a good deal of modification to (working) BASIC programs may be required before they will compile successfully. However, compilers are available for a number of the best selling micros (eg. Commodore, Apple, BBC etc.) and for CP/M (see below) users, who have access to the Microsoft range of BASIC compilers such as CBASIC and MBASIC (the latter being compatible with the MBASIC interpreter). For the PC user Microsoft's QuickBasic and Borland's Turbo BASIC provide very easy-to-use compilers which are largely compatible with the PC's own BASIC (or BASICA on a clone), while at the same time offering a significant increase in execution speed and many advanced language features not normally associated with BASIC (such as named subroutines, sophisticated event trapping, and improved data structures). QuickBasic also provides a good and built-in full screen editor for preparing the program before compilation, can support very large user programs and has facilities for calling machine code routines and for passing parameters between such routines and the calling program.

Also between BASIC and assembler lie the alternative high-level languages, most imported from the world of mainframe computers. In the context of laboratory instrumentation the C language may be regarded as particularly useful. C programs are much easier to write than assembler programs, because all of the arithmetic and logical constructions are available (rather like BASIC, but more comprehensive), including floating point arithmetic. However, C programs can access all parts of the computer (including interface connections) that can be reached with machine code, and the compiled program runs almost as fast as assembler. Furthermore, C compilers are available for most modern microcomputers using 68000 or 8086 family processors, so it becomes relatively straightforward to transfer programs from one to another. It is only relatively straightforward, as, for example, when transferring between an Amiga and a PC, some changes are likely to be required to the keyboard reading and screen display functions.

It is probably worth pointing out that many high-level languages (FORTRAN, Pascal,

Algol, etc) derive from mainframe computer systems, where they are primarily used for programs which receive character input and generate character output. To use such languages for preparing programs which generate multicolour graphics or window displays on a micro is not quite as straightforward as it is with BASIC. Generally the use of the micro's display facilities is possible only through the use of subprograms designed specifically for the task - and such subprograms must be provided either by the user or by purchasing libraries of subprograms. At the time of writing there are numerous libraries on the market (from companies such as Microsoft and Digital Research), although the costs are significant. Some of the "newer" languages (such as Turbo Prolog from Borland and Microsoft's Quick-C) are supplied with excellent libraries of routines for using colour graphics and window displays.

Finally it should be noted that micros are increasingly being provided with facilities for incorporating math coprocessors, which are essentially second microprocessors specifically designed for performing floating point arithmetic on decimal numbers with many significant digits. For example, the 8087 and 80287 math coprocessors can be incorporated into PCs using the 8086 and 80286 MPUs respectively, and these coprocessors can perform floating point multiplications directly using an IEEE standard representation of floating point numbers, while the MPUs can only perform integer arithmetic. Many compilers (but not all) offer the ability to generate code which can use a coprocessor, and such code will run considerably faster than code which has to handle arithmetic by converting floating point numbers to integers and back again. In a typical scientific calculation program (ie. where most of the arithmetic is floating point) the use of the coprocessor can lead to a speed increase of a factor of 10 - but only if the program contains the correct instructions to activate the coprocessor. Microsoft's QuickBASIC and Borland's Turbo BASIC are two BASIC compilers which offer considerable refinements to the BASIC language and could be regarded as competitive products, and both are supplied with the facilities to make use of a coprocessor. For other languages machine code libraries to handle coprocessors are usually available from coprocessor manufacturers such as Hauppage Computer Works.

5.8 The operating system

While microcomputers are programmed in popular languages such as BASIC, they are all capable of carrying out operations which are not part of the high level programming language. Loading and saving programs on disk, editing programs or files and printing characters on the screen or printer are examples of such functions. These aspects of a micro's operation are handled by a special program called the operating system. In some cases the operating system is almost invisble to the user (and referred to as "transparent") - in fact it is possible to use some micros without being aware of any distinction between the operating system and the principal programming language. However, moving from one micro to another can make one uncomfortably aware of such a distinction, particularly when the new operating system is both non-transparent and relatively unfriendly.

The appearance of an operating system to the user (ie. what you actually have to type on the computer's keyboard to make the machine do something) varies widely between micros. Many have a system which is unique to that model of computer, while others have adopted one of the major standard operating systems, such as CP/M (which may stand for Control Program for Microcomputers, although opinions about this differ) for 8-bit micros, or MS-

DOS or CP/M-86 for 16-bit machines or UNIX (or one of its many variants) for 32-bit machines. These standardised operating systems are normally loaded from disk and used in RAM, as are the various interpreters, compilers and applications programs generally used under the operating system.

Until recently the computers most likely to be encountered in a laboratory context tended to have unique operating systems, so there was little to be gained from considering any one particular system. However, the advent of the PC and the newer 32-bit machines has changed that situation, and it is now well worth a closer look at one of the major systems. MS-DOS is considered in some detail in the next chapter. A good operating system can make an enormous difference to the ease with which a laboratory system can be programmed, tested and operated, but, much more importantly, it can also dramatically affect the range and quality of the software available for a specific microcomputer. Thus an enormous number of programs is available, both commercially and via exchanges, for operation under the MS-DOS operating system - exceeding the sum of all those available for the CP/M86 and Xenix systems for the PC and those for the Amiga, Commodore, Macintosh unique operating systems.

In choosing a microcomputer for relatively fast applications (eg. where thousands of byte transfers per second are envisaged), or for applications which are likely to require the manipulation of thousands of bytes, it is wise to ensure that the micro's operating system is well documented. This allows the user to find the information required to use the system's built-in machine code routines for some aspects of data handling, rather than writing his own in machine code or relying on relatively slow BASIC for everything. For some systems this information is widely available (eg. for MS-DOS in "Advanced MS-DOS" or "Programmer's guide to the IBM PC," or for AmigaDOS in the Commodore manuals). For others the documentation is often exchanged through users' clubs, and any serious micro user is bound to benefit from joining one of these. Until recently the value of documentation provided by the micro's manufacturer has ranged from fair to poor.

5.9 Byte transfers

The details of the operation of the MPU within a microcomputer are beyond the scope of this text; several sources of such information are included in the bibliography. However, some aspects of interfacing (dealt with in chapters 7 and 8) require a somewhat deeper appreciation of the control of byte transfers over the data bus than has been necessary so far. Here we introduce in outline the manner in which the MPU reads bytes from, and writes bytes to, a specified address in RAM, as a prelude to our later discussion of reading and writing bytes to or from circuits associated with laboratory interfaces.

Every operation of the MPU is carried out on a specified transition of its clock input, ie. when the level on the clock input is changing from hi to lo or lo to hi. Consequently, collecting addresses from the address bus and placing data bytes on, or collecting data bytes from, the data bus requires that other circuits on the bus (eg. ROM, RAM, PIA) are synchronised with the MPU. This synchronisation is achieved by signals on the control bus. The precise details of the control bus signals vary with the MPU, but most fall into one of the two main categories, illustrated in part in figs 5.6 and 5.7.

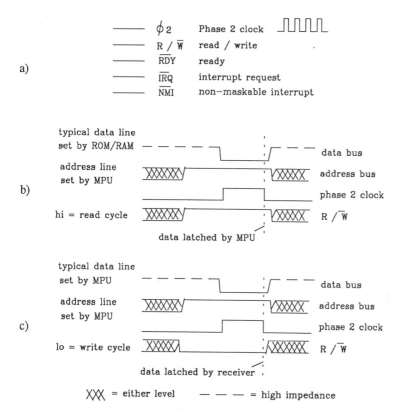

a)

⎯⎯⎯	$\phi 2$	Phase 2 clock ⎍⎍⎍⎍
⎯⎯⎯	R / W̄	read / write
⎯⎯⎯	R̄D̄Ȳ	ready
⎯⎯⎯	ĪR̄Q̄	interrupt request
⎯⎯⎯	N̄M̄Ī	non−maskable interrupt

b)

typical data line
set by ROM/RAM ⎯ ⎯ ⎯ ⎯ ⎯⎤ ⎡ ⎯ ⎯ data bus

address line
set by MPU ⊠⊠⊠⊠⟋ ⟍⊠⊠⊠ address bus

⎯⎯⎯⎯⎤ ⎡⎯⎯⎯ phase 2 clock

hi = read cycle ⊠⊠⊠⊠⟋ ⟍⊠⊠⊠ R / W̄

data latched by MPU

c)

typical data line
set by MPU ⎯ ⎯ ⎯ ⎯⎤ ⎡⎯ ⎯ ⎯ data bus

address line
set by MPU ⊠⊠⊠⊠⟋ ⟍⊠⊠⊠ address bus

⎯⎯⎯⎯⎤ ⎡⎯⎯⎯ phase 2 clock

lo = write cycle ⊠⊠⊠⟍_____⟋⊠⊠⊠⊠ R / W̄

data latched by receiver

⊠⊠⊠ = either level   ⎯ ⎯ ⎯ ⎯ = high impedance

Fig 5.6 *a) Some of the control bus lines used by MPUs which control their read/write operations through a R/W line and a clocking signal. The synchronisation of the b) read, and c) write operations. The cross hatching indicates that the levels on the lines are irrelevant.*

In fig 5.6a we see the principal control bus lines and signals of MPUs such as the 6502 and 6800. Operation may be understood by examining the "read byte from memory" and "write byte to memory" sequences illustrated in fig 5.6b and c respectively. The data bus is a TRI-STATE bus and is normally in the high impedance state. When a byte is to be read from memory the MPU places the byte's address on the address bus and holds the read/write (R/W) control bus line hi to signify a read operation (ie. the addressed device is expected to place the data byte on the data bus). Now nothing actually happens anywhere else until a second control bus line (called the phase 2 clock) goes hi. Phase 2 is actually a clocking signal closely related to the MPU's clock input, although sometimes it is a non-symmetrical signal. When phase 2 goes hi the addressed device (say a ROM) is supposed to respond by placing the value of the addressed byte onto the data bus. As this process can require one or two hundred nanoseconds before the data bus lines stabilise to the necessary his and los, the MPU waits until phase 2 returns to lo. This hi to lo transition corresponds to the instant when the MPU latches the data from the data bus into one of its own buffers or registers (ie. it has read the data byte), and at this point the address and data bus lines and R/W are freed for use in the next instruction (the data lines, being TRI-STATE, return to the high impedance state).

A similar sequence of events governs the writing of a byte to memory, as illustrated in fig 5.6c. The MPU sets the address lines to the required values and sets R/W lo to indicate a write operation. The rise of phase 2 from lo to hi coincides with the MPU placing a data byte on the data bus. That data byte is stable by the time that the phase 2 line returns to lo, and this hi to lo transition is used by the addressed device to latch the data line values into RAM or the PIA. Shortly after the transition the MPU releases the address lines and R/W, and the data lines become high impedance. (Note: the situation is slightly more complex in the case of the 6800 because a "valid memory address," VMA, control line is also used.)

The key feature of the technique illustrated in fig 5.6 is that a R/W control line is used to indicate whether a read or write operation is required, and the phase 2 clock pulse edges are used to provide the timing signals. The other major read/write technique uses one control line to control read operations and another control line to control write operations. This approach is illustrated in fig 5.7, and is used by the 8080 MPU and (in a modified form) by the Z80. Again the MPU places the address information on the address bus, but now either the normally hi MEMR (memory read) is taken lo to indicate a read operation (fig 5.7b), or a normally hi MEMW (memory write) line is taken lo to indicate a write operation (fig 5.7c). In each case the latching of the data, either by the MPU or by RAM, occurs when the MEMR or MEMW undergoes a lo to hi transition.

Of particular interest with some MPUs using the second technique (fig 5.7) is the availability of a second pair of control lines IOR (IO read) and IOW (IO write) which can be used to control byte transfers between the MPU and an alternative set of addressed locations when specific instructions are carried out. Thus while normal load and store instructions cause addresses to be loaded and the MEMW or MEMR lines to be toggled, instructions such as IN and OUT cause addresses to be loaded but the IOW or IOR lines to be toggled while the MEMx lines remain hi. The Z80 addressing method is similar but uses one pair of control lines for RD and WR, and a second pair to distinguish between memory access and IO access - MEMRQ and IORQ. A consequence of these arrangements is that a micro with a 16 bit address bus and a full complement of 64k bytes of memory may be programmed to carry out byte transfers with the 64k of memory *and* with up to 64k of addresses outside memory - a potential which has to some extent been nullified by certain hardware features in some low-cost Z80-based micros. An essentially similar technique is adopted for byte transfers involving the 8086 family of processors, and we shall return to this subject in chapter 7.

The popular 8-bit MPUs are provided with a 16 bit address bus and an 8 bit data bus, so that 24 pins of the MPU IC are required for byte transfers with an addressing range of 64k, plus of course the control signals, clock and power supply pins. Until recently it was not possible to produce memory ICs containing such a large number of addressable bytes; most of the memory ICs held 4k, 8k or 16k bytes. Consequently the memory of most of the popular microcomputers was made up of several separate ICs, some ROM and some RAM. For this reason it was not the practice to connect all 16 address lines to every memory IC, but rather to partially decode the 16 bit address into two parts. Typically the lower 12 bits of the bus (covering the range 0 - 4096) were left as address lines, while the top 4 bits were decoded to provide 16 separate signals, as illustrated in fig 5.8. Each of these could be used to identify which one of up to 16 blocks of 4k bytes should react to the address held on the lower 12 bits of the address bus. This technique was widely used for addressing ROM, allowing standard

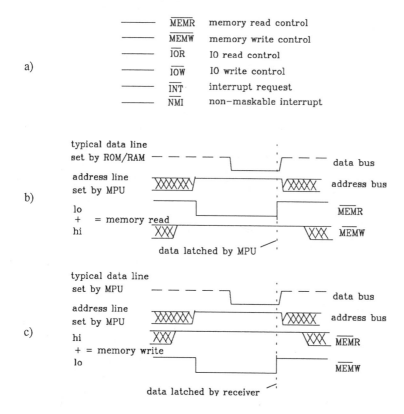

Fig 5.7   a) *Some of the control bus lines used by MPUs which control their read/write operations through separate MEMR and MEMW lines. The synchronisation of the b) read, and c) write operations. The cross hatching indicates that the levels on the lines are irrelevant.*

4k ROMs (in 24 pin packages) to be used in systems containing >20k of ROM.

Most microcomputers have facilities allowing relatively straightforward connection to their address, data and control busses, often provided for the purpose of allowing memory expansion. In some cases (eg. the Commodores and Amigas) a connector is available on the case of the micro. In others the connector is provided inside the case with no obvious hole for a cable, or with clearly indicated slots to allow connection of additional circuits (eg. the I/O slots of the Apple II and PCs). On the Apple while all 16 address line are provided on the I/O slot connectors, partially decoded address signals also appear on two of the connector pins, one for a given value of A4 - A15 (leaving A0 - A3 for use), and one for a given value of A8 - A15 (leaving A0 - A7). The BBC microcomputer provides a 1 MHz Extension Bus connector on which address lines A0-A7 appear, while the data bus is gated by block select lines which also appear on the connector and provide a communication pathway using address blocks $fc00 (known as FRED) and $fd00 (JIM). Our only interest in connection to the address, data and control busses is for interfacing purposes, and for that reason we defer further discussion until the latter part of chapter 7.

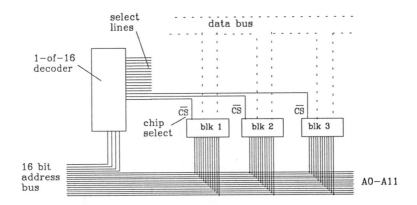

Fig 5.8 *Partial decoding of a 16 bit address bus to provide a 12 bit bus and 16 separate block select lines. In this example each ROM block decodes the 12 bit address on A0-A11 when its chip select line (CS) is held lo.*

5.10 External byte transfers

Most microcomputers have additional systems connected to their three busses, and the most important in the present context is the peripheral interface adapter (PIA) circuit, (appearing sometimes as peripheral input/output (PIO) or programmable peripheral interface (PPI), or as the more sophisticated versatile interface adapter (VIA) or complex interface adapter (CIA) circuits). PIA circuits are used to allow the MPU to communicate with other byte handling systems which are not synchronised with the computer's clock and which could give rise to timing difficulties if the data bus itself was used as the communication channel. The keyboard, printers and "user ports" are examples of systems often connected through PIA circuits.

While a real PIA system is quite complex, for our present purposes it is sufficient to regard it as a circuit which allows electrical connections with the computer's byte handling data bus, and our interest will concentrate on the use of a PIA in forming parallel user-ports. These usually consist of a multi-pin connector which can carry the eight parallel digital signal levels which make up an 8-bit byte of data, and one or two other signals used for control purposes. A typical example is illustrated simply in fig 5.9, although we shall examine a parallel port system in more detail in the next chapter. The advantage of such a system is that we can input the byte value of eight hi or lo logic level signals present on the port wires, or output eight hi or lo logic levels, using a BASIC program. Some computers use specific instructions (such as INP and OUT) to transfer bytes in and out of parallel ports, while others treat the PIA circuit as though it was a byte of memory (hence the term "memory-mapped" port) and we may utilise the PEEK and PQKE type of instructions we met in addressing memory bytes. As single IC PIA and VIA circuits compatible with most of the popular MPUs are manufactured, micros which are not fitted with a (useable) PIA can generally have one fitted with little difficulty. In some cases "plug-in" boards containing one or more PIA circuits are available for connection to some convenient point on the micro's circuit board (eg. the expansion slots in the Apple II and IBM PC).

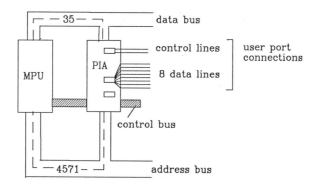

Fig 5.9 *Some useful connections on the user ports provided on the Amiga, BBC and Commodore microcomputers.*

While it is possible for a micro to send bytes to or receive bytes from a peripheral through a PIA port without any additional signalling, such an approach would be prone to failure unless the micro and the peripheral were operating at precisely the same speed. In practice such a synchronisation of devices is difficult to achieve. Some devices connected to a computer may only be able to handle data bytes much more slowly than the computer can despatch them, even when under the control of a BASIC program. Low cost, unbuffered printers are an obvious example, being able to print at perhaps 80 characters per second while the computer can transmit several thousands of bytes during the same period. For this reason alternative techniques are employed to ensure that one device transmits a byte only when the other device is able to receive it. One simple approach may be found in using two control lines, one carrying a signal from the computer to the peripheral, and a second carrying a signal in the other direction. The basic principle is illustrated in fig 5.10, where the Cx1 and Cx2 control lines of a typical PIA port are used to carry these "handshaking" signals to a simple byte handling peripheral.

We will discuss the byte transfer process used for output of the data byte first, because many micros are equipped with an output port which utilises these principles, often for connection to a printer. (The Centronics[R] parallel interface is closely related to this system but requires open collector data bus drivers.) The signal levels used during the transfer are included in fig 5.10. Firstly we should note that the PIA in this example is programmed so that its Cx2 control line may be set either hi or lo, and so that it responds to a lo-to-hi transition on its Cx1 control line by setting a flag (ie one bit of a register) to a 1. Of course, it would be equally valid to program the PIA so that it responded to a hi-to-lo transition if the peripheral's hardware gave signals complementary to those illustrated in fig 5.10. When the computer wishes to "handshake" a data byte to the interface the following sequence occurs:

1. The computer places a byte on the PIA's port data lines,
2. The computer holds the Cx2 control line lo to signify that the data on the data lines is valid, and then waits for the PIA flag to be set.
3. The peripheral holds the Cx1 line lo, to indicate that it is busy digesting the data on the data lines.

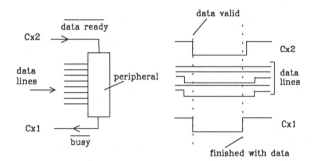

Fig 5.10 *A two wire "handshaking on output" technique. The peripheral (left) generates a busy signal until it has finished with the data. The control line signal levels are illustrated on the right.*

4. When the peripheral has digested the data byte it returns the Cx1 line hi, which sets the flag in the computer's PIA.
5. The computer observes that the PIA's flag has been set. It clears the flag and returns Cx2 hi (to indicate that there is no valid data on the data lines).
6. If any more data are to be sent the cycle starts again at step 1.

For byte input the situation is slightly different as it is the interface or peripheral which must place the data byte on the data bus and signal to the computer that the data bus is ready for reading. An example of data input is given in fig 5.11 for a peripheral handshaking data bytes to the computer. In this example the following sequence occurs:

1. The computer drops the Cx2 line from its normally hi level to lo, which informs the peripheral that it is ready to receive data, and waits for the PIA's flag to be set.
2. The peripheral sets the Cx1 line lo to indicate that it is preparing the data but that the data bus is not ready.
3. When the data is ready and loaded on the data lines and stable, Cx1 is returned hi by the peripheral. This sets the PIA's flag in the computer.
4. The computer finds that the PIA flag has been set. The computer then clears the flag, reads the data bus and returns the Cx2 line hi.
5. If another transfer is required the cycle starts again at step 1.

Note that in fig. 5.11 the individual data lines have their level set by the peripheral at different times. Clearly attempting to read the data lines too early (before the busy line goes high) would result in the transfer of an incorrect data byte.

The techniques for byte input and output with handshaking are very reliable and moderately fast (typically up to 10000 bytes a second may be transferred under the control of a machine code program) if neither the computer nor the external device imposes significant limitations. Parallel user ports, which allow byte input or output with handshaking, provide one of the most useful and inexpensive connections for byte transfer between micros and laboratory instrumentation. Details of a simple user port adaptor for the IBM PC are

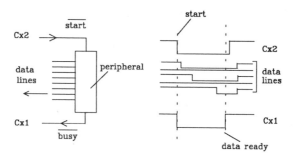

Fig 5.11 *A two wire "handshaking on input" technique, illustrated with a peripheral which generates a busy signal until it has set up the eight data bits and these are stable on the data lines. The control line signal levels are illustrated on the right.*

discussed in the next chapter, and the techniques for programming the byte transfers are outlined in chapter 8.

A second connection system provided on many micros is a standard input/output interface connection, some types of which we shall be discussing in chapter 8. Popular connection systems include the IEEE 488 standard system, the Centronics[R] parallel standard port (which provides output only) and the RS232C-type standard system (appearing as RS232C or, in a modified form, as RS423). With the exception of the IEEE 488 system, these connections are provided primarily to enable the micro to be connected to computer peripherals, such as printers, plotters and modems, rather than laboratory instrumentation. Nevertheless the RS232C standard is becoming increasingly popular as a standard connection system for laboratory instruments, and so is providing an increasingly used route for the transfer of data between such instruments and a microcomputer. The principal advantage of these kinds of connections is that their use from within a BASIC program may be much simpler than the alternative, less standardised interfacing systems we meet in chapter 7. The price paid for such convenience is generally a cash one, but if a system is complex and expensive the additional cost of a standard I/O interface is probably worth paying.

5.11 Interrupts and interrupt flags

The control bus of all MPUs includes at least one line which can be used to interrupt the execution of a program (see for example figs. 5.6a and 5.7a). Many MPUs provide two different interrupt lines, although microcomputers based on these MPUs do not necessarily have both of these lines connected to anything. An interrupt is signalled by some device external to the MPU taking the interrupt line lo. When this happens the MPU stops execution of its program, stores the program counter register (which holds the address of the next instruction it would have carried out if it had not received the interrupt) and the status register in a part of RAM known as the stack, and then starts executing a second program. The start address of this second program is known as the interrupt vector, and is stored at a specific memory address which is normally part of ROM. For example, the interrupt request vector for the 6502 is expected to be at addresses 65534-5, $fffe-f in hexadecimal.

Microcomputers tend to rely heavily on the use of interrupts to carry out fundamental operations, such as the checking of the keyboard to discover which key has been pressed. Some are wired up so that an interrupt occurs when specific external events take place (such as a key being pressed), others arrange for frequent and regular interrupts (such as 50/60 times a second). All have an interrupt servicing routine as part of their operating system, and this is pointed to by the interrupt vector - the program address to which control is passed whenever an interrupt occurs. The interrupt servicing routine generally stores pointers and data associated with the program which was being executed, and then examines a number of parts of the system to see what caused the interrupt and to take whatever action is required. Once the interrupt has been serviced, the system restores the data and pointers of the original program and returns to execute that program at the point of the interruption. (For most microprocessors a special return instruction, RTI, causes the MPU to reload its saved program counter and status registers from the stack. Note that many microcomputers use two stacks - one for the MPU, and a second for the interrupt servicing routine of the operating system.)

Those MPUs with two interrupt lines (eg. the 6502, 6800, Z80 etc.) generally allow the user to prevent or mask an interrupt on one of the lines (the INT or IRQ line). This facility allows, for example, the interrupt servicing routine to ensure that there will not be a second interrupt before it has finished servicing a first interrupt (otherwise things could get very complicated). However, the second interrupt line is not maskable (it is called the NMI line - Non-Maskable Interrupt), and an interrupt on this line is always attended to by a second servicing routine with an address given by the NMI vector, and which probably has to be supplied by the user. (The 6502 NMI vector is at 65530-1, $fffa-b.) Considerable care is needed in the use of NMIs, although if they are not used by the micro's operating system they do offer a straightforward way of connecting external devices which need attention on demand.

MPU interrupts are not easy to handle without a good deal of practice and assembler language programming, although their value for detecting external events (eg. requests for attention from laboratory instruments) is considerable. Because this book is intended primarily for the high level language user we shall not deal further with the use of MPU interrupts. Fortunately micros with PIA circuits usually possess the facilities for setting a one bit flag (that's just a bit of a byte) when a logic level transition is detected on a PIA control line. This allows the user to program a response to external events very much more easily, although less efficiently, than through MPU interrupts.

CHAPTER 6

THE PC FAMILY

In essence the PC is a type of microcomputer, originally created by IBM in 1981 and more recently produced in a variety of forms by dozens of manufacturers throughout the world. The original components of the PC have been upgraded several times during the last few years, so that while the early models may have had an 8088 processor, 128k RAM and one low capacity floppy disk, later models are more likely to have an 80286 processor, 640k RAM, a high density floppy disk and a 20 Mbyte hard disk (the normal PC/AT system). However, the apparent structure of these variants (although not necessarily the actual structure) has remained remarkably compatible over the years, and the PCs remain basic micros which can run a relatively standardised operating system and can communicate with external devices using a relatively standardised but versatile connection system.

So much has been written about the PC that there would be no purpose in repeating a detailed description of the machine here. The excellent books on hardware (by Sargent and Shoemaker) and programming (by Norton) provide detailed and highly readable accounts of these aspects of the PC. The outline provided below is primarily intended to emphasis those aspects of the machine specifically related to connection with external devices.

6.1 The elements of a PC

While the actual appearance of the PC variants differ in detail, the inside of a typical PC usually resembles the diagram shown in fig 6.1. Most PCs are of modular construction, so that components such as the power supply, or a disk drive, or even the system board (which holds the processor and control circuits) can be removed relatively easily. From our point of view the power supply is a relatively important component, in that interfacing circuits plugged into the PC's expansion slots may be powered by the internal power supply, and indeed may provide limited amounts of power for external devices. For this reason it is important to bear in mind that a power supply is more than just a power supply; it is a power supply capable of supplying up to a specified amount of power on each of it supply lines (usually +12, +5, -5 and -12V). Power supplies fitted to PCs range in their power rating from about 130W (for the early PCs) to over 250W for some PC/AT clones.

The power supply connects to the main system board which is secured to the base of the cabinet. Above the system board there is usually room for the floppy and hard disk drives held in a supporting cage. There is an unfortunate lack of standardisation in the positioning of supports and securing holes for these peripherals, and in the half dozen makes of PC in our

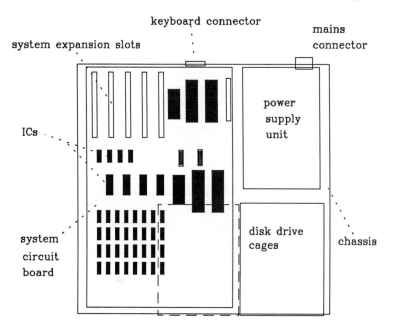

Fig 6.1 *The major components of a PC system unit*

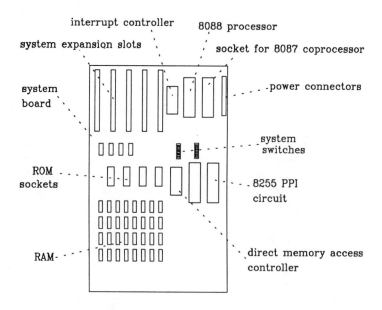

Fig 6.2   *Layout of the system board of a typical PC clone, showing the major elements of the circuitry and connectors.*

Table 6.1 The PC's address space

Address range	Function
0 - 3FF	Interrupt vectors
400 - 47F	BIOS RAM
480 - 5FF	System function RAM (eg. BASIC)
600 - 9FFFF	Program memory (Base memory)
0A0000 - 0AFFFF	IBM enhanced graphics adaptor
0B0000 - 0B0FFF	IBM monochrome adaptor
0B8000 - 0BFFFF	IBM colour graphics adaptor
0C0000 - 0CFFFF	Reserved for I/O
0D0000 - 0EFFFF	ROM cartridges
0F0000 - 0F3FFF	Unused
0F4000 - 0F5FFF	Spare ROM sockets
0F6000 - 0FDFFF	BASIC ROM
0FE000 - 0FFFFF	BIOS ROM

laboratories a wide range of adaptors and wedges have been found necessary to secure such devices. Most modern floppy disk units are "half-height" (ie. they occupy half the height of the cage, so that one can be installed above another), as are many of the medium capacity hard disks (< 30 Mbyte), while high capacity hard disks are often "full-height". A typical cage can hold four half-height devices (including tape streamers and removable hard disks), although some clones offer space for a fifth.

A typical system board has a layout similar to that shown in fig 6.2 and contains many of the PC's electronic circuits, including the processor and its supporting chips, a socket for a floating-point co-processor, at least part of the RAM, and chips which handle the keyboard and loudspeaker. The address space of the processor in the standard IBM PC is shown in Table 6.1. The PC compatibles are constructed around 8088 or 8086 processors which have an address range limited to 20 bits (640k, the range 0 - 0FFFFFH). Addressing additional memory requires that the additional RAM shares addresses with other components from Table 6.1. Three companies (Lotus, Intel and Microsoft) have agreed on a standard approach to swapping blocks of additional memory for blocks of base memory (the LIM standard), and a number of add-on memory cards are available which allow an expansion of the PC's memory using the LIM standard techniques. Additional memory which conforms to the LIM standard is called Expanded memory. Many large application programs are able to make use of such expanded memory to use more than 640k of RAM.

PC/AT compatibles and micros based on the 80386 may generate addresses above 0FFFFFH, and are often equipped with more than 640k of RAM. Memory which is mapped at addresses above 1000000H (1 Mbyte) is referred to as Extended memory. (It is a pity they both begin with E; this has been known to cause confusion.) Some operating systems and supervisory software are able to use extended memory in addition to base memory (although MS-DOS up to version 3.3 cannot). For example Windows-386 uses available memory to create several blocks of up to 640k each within which applications are run as though in an otherwise empty PC. More mundane software allows RAM disks (ie areas of memory pretending to be very fast disk drives) and disk caches (ie large and very fast buffers which reduce the number of times a disk drive actually needs to read a disk) to be set up in extended memory. Memory management software is also available for many machines to allow extended memory to behave as LIM standard expanded memory - so permitting applications which support the LIM standard to use more than 640k.

At the rear of the system board is a collection of edge connectors (slots) which provide access to the system expansion bus. Typically between 5 and 8 connectors are provided, although in most working PC systems at least some of these slots will be occupied by circuit cards which perform the functions of controlling disks and video display systems. In most cases the circuit cards are fixed inside the cabinet running from front to rear, although in the case of the Amstrad PC1640 the cards run from one side to the other. Those cards which make connections to internal devices (such as disk drives) have connectors which allow flying leads to be passed to the device, while those intended for external connection (to, say, printers, video displays and laboratory instruments) usually provide sockets which protrude from the PC's cabinet.

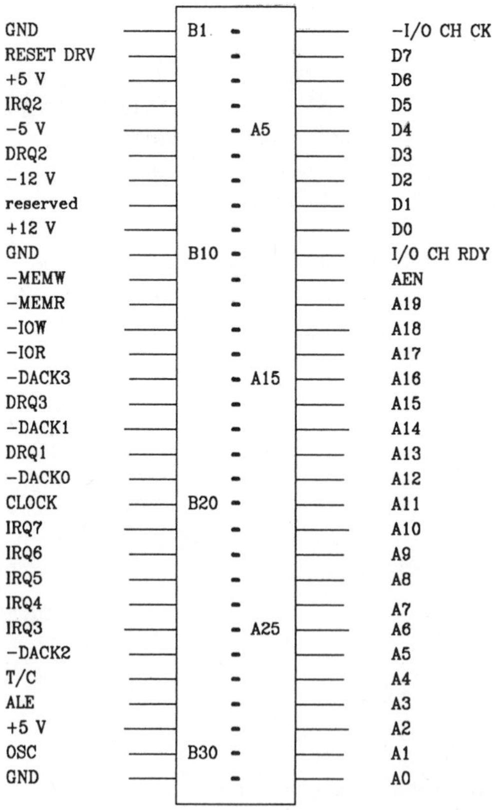

GND	B1	−I/O CH CK
RESET DRV		D7
+5 V		D6
IRQ2		D5
−5 V	A5	D4
DRQ2		D3
−12 V		D2
reserved		D1
+12 V		D0
GND	B10	I/O CH RDY
−MEMW		AEN
−MEMR		A19
−IOW		A18
−IOR		A17
−DACK3	A15	A16
DRQ3		A15
−DACK1		A14
DRQ1		A13
−DACK0		A12
CLOCK	B20	A11
IRQ7		A10
IRQ6		A9
IRQ5		A8
IRQ4		A7
IRQ3	A25	A6
−DACK2		A5
T/C		A4
ALE		A3
+5 V		A2
OSC	B30	A1
GND		A0

Fig 6.3 *The connections provided on the PC's expansion bus slots.*

The expansion bus slots in PCs (at least the original PC, the PC/XT and clones) are all based on a 62-way edge connector with the assignments detailed in fig 6.3. The connections essentially allow access to the PC's 8-bit wide data bus and 20-bit wide address bus. However, the PC/AT introduced additional 36-way connectors positioned directly in front of some of the standard PC connectors (shown in fig 6.2) and carrying the extra lines needed to provide access to a 16-bit wide data bus and 24-bit wide address bus. All the 62-way connectors are identical, but some cards specifically intended for the AT have mating edge connections which join both the 62-way slot and the smaller 36-way AT slot, and so will not work if connected to only the 62-way slot. (Some cards intended for the PC/AT are about half an inch taller than will actually fit inside a PC or PC/XT with the cabinet in position, even if they don't require the AT connector. Furthermore, some cards intended for the PC/XT will not fit into an AT slot which has the additional 36-way connector in front of it, because the card itself clashes with the additional connector!)

Table 6.2   IO address map for the IBM PC/XT

Hex address range	Device
000-01F	DMA controller
020-03F	Interrupt controller
040-05F	Timer
060-06F	PPI
080-09F	DMA page registers
0A0-0AF	NMI Mask registers
200-20F	Game control
210-217	Expansion unit
21F	Reserved
278-27F	Parallel printer port
22B0-2DF	Alternate Enhanced Graphics Adaptor
2E1	GPIB (adaptor 0)
2E2-2E3	Data Acquisition (adaptor 0)
2F8-2FF	Serial port 2
300-31F	Prototype card (eg user port adaptor, see below)
320-32F	Hard disk units
348-357	DCA 3278
360-36F	PC Network
380-38F	SDLC communication
380-38F	Binary synchronous communications 2
390-393	Cluster
3A0-3AF	Binary synchronous communications 1
3B0-3BF	Monochrome display and printer port 1
3C0-3CF	Enhanced Graphics Adaptor
3D0-3DF	Colour Graphics Adaptor
3F0-3F7	Floppy disk controller
3F8-3FF	Serial port 1

The expansion slots are of interest for two reasons. Firstly they hold the cards which provide quite fundamental features of the computer system, such as the display adaptor which generates the signals for the monitor display, controller cards which handle byte transfers between RAM and disk, and standard interfaces for connection to printers, modems, mice and other peripherals. Secondly they can hold special cards, including "in-house" cards, for interface connection to laboratory instrumentation. The connections of the 62-way expansion slot are shown in fig 6.3 and it may be seen that an 8 bit data bus and 20 bit address bus are provided, along with a number of control bus signals. The busses are not necessarily connected directly to the processor (the 8088, for example, has only 16 pins available for addresses, so it generates 20 bit addresses by using some pins to carry two different address bits at different times), although for our purposes we can regard the expansion bus as containing the 20 bit address and 8 bit data signals and the control signals necessary to allow transfers through the expansion bus to be performed.

Cards connected to the expansion bus may be exchange bytes with the rest of the system either using memory instructions (ie X=PEEK(1234) or mov ax, 1234) or using IO instructions (ie X=INP(1234) or in ax,(1234)), provided that the memory or IO address does not conflict with an address on another card or within the PC. [Cards may also transfer data using Direct Memory Access (DMA) techniques, in which the address bus is loaded by card circuits rather than by MPU instructions. However, the these techniques are beyond the scope of the present text.] The Technical Reference Manuals for most PCs specify the address ranges, both memory and IO, which are reserved for specific purposes. The memory addresses used in a standard PC are specified in Table 6.1, while the IO addresses reserved for IBM approved devices are listed in Table 6.2.

6.2 Video display adaptors

One of the expansion slots of almost every PC will be occupied by a card which generates the signals required to operate a display device, such as a video monitor. The card usually contains a connector which protrudes from the PC's cabinet, allowing a multiwire cable to be plugged in to carry the signals to a monitor. The card is called a video display adaptor.

There are five basic types of video display adaptor available for the PC: the IBM monochrome adaptor, the Hercules (Hercules Computer Technology) monochrome graphics adaptor, the IBM colour graphics adaptor (CGA), the IBM enhanced graphics adaptor (EGA), and the IBM professional graphics adaptor (PGA). All of these offer a text mode of display, so utilise the character generator technique described in chapter 5, and most offer one or more graphics display modes using pixel-mapped graphics. To implement these displays all the adaptors contain video RAM which is quite independent of the PC's RAM, although naturally mapped into the address space of the PC's processor. Communication between the PC's processor and the adaptor's RAM is carried out in the same way as communication between the processor and its normal RAM, although in this case via the address, data and control busses on the expansion slot. However, the adaptor's RAM is also examined by logic circuits on the adaptor card which interprets the RAM contents to produce the video signals which generate the required display. The adaptor's RAM is thus accessed by both the PC's processor and the adaptor's logic circuitry (at different times), and is referred to as "dual port" RAM. The adaptors all contain different amounts of RAM starting at different address, although in all cases the first address is chosen so that it cannot conflict with a RAM address in the PC's memory.

The adaptors also contain a number of registers which are used to control aspects of the display (such as the resolution, the position of a cursor or the number of characters per line, etc.) or to provide information which allows PC software to synchronise actions with the display. 8 bit patterns may be directed to or derived from these registers using sequences of instructions based on IN and OUT (the assembler equivalent of INP and OUT) using a small number of port addresses. Accessing the adaptor's registers should not be undertaken lightly, as it is possible to burn out the display device by improperly setting some parameters.

The IBM monochrome adaptor provides a single display mode of characters only, consisting of 25 lines of 80 characters per line. The adaptor contains 4k of RAM occupying addresses 0B0000H - 0B0FFFH and uses 18 registers for control and synchronisation purposes with port addresses 3B0H - 3BFH (some of which are used for a parallel printer port provided on the display adaptor). The adaptor uses two bytes for every character position of the display arranged as adjacent pairs. The first byte of each pair holds the character code representing the letter/number/symbol to be displayed, while the second byte holds the "attribute" information which allows the adaptor to display each character with its own attributes as normal, reversed, highlighted or underlined. The character displayed is generated from the character code held in RAM through a character generator system formed from a ROM containing the display patterns, and characters are formed from a 9*14 dot matrix.

The Hercules monographics adaptor provides the facilities of the IBM monochrome adaptor (including the parallel printer port) and the ability to display bit-mapped graphics using 720 (horizontal) * 348 (vertical) pixels. The adaptor contains 64k of RAM, although only 32k is required for a single bit-mapped display. This allows two different displays to be stored and used - although obviously only one may appear on the display device at any one time. A later adaptor from Hercules (the Graphics Plus card) includes a selectable character generator which permits a variety of different character sets to be displayed on screen at the same time. Hercules cards are usually supplied complete with a version of BASIC (HBASIC) which allows BASIC commands on IBM computers to generate monochrome graphics displays. The resulting ease with which good resolution displays can be generated makes the Hercules card a very attractive proposition. The standard BASICA and GW-BASIC of PC compatibles support only CGA graphics, although the Hercules Graphics Toolkit from Laboratory Software offers a comprehensive set of graphics routines (including graphic screen dump routines) which may be called from a variety of languages including BASIC.

The monochrome adaptors generate signals for a 350 line raster scan at a frequency of 18.5 kHz (ie horizontal lines per second). These signals are intended to drive specific monochrome video displays. Because the flyback of the electron beam is used to generate the high voltage bias of a video display it is important that it is driven at an appropriate line rate. Using an incorrect line rate can result in damage to the monitor or to the driving display adaptor.

The IBM Colour Graphics Adaptor (CGA) permits seven different display modes, four of

Table 6.3   Display modes available with the CGA

mode	display	RAM	colours[a]	DOS mode	BASIC commands
0	40 * 25 text[b]	2k	16 greys	BW40	SCREEN 0,0:WIDTH 40
1	40 * 25 text	2k	16/8[c]	CO40	SCREEN 0,1:WIDTH 40
2	80 * 25 text	4k	16 greys	BW80	SCREEN 0,0:WIDTH 80
3	80 * 25 text	4k	16/8	CO80	SCREEN 0,1:WIDTH 80
4	320 * 200 pels[d]	16k	4	-	SCREEN 1,0
5	320 * 200 pels	16k	4 greys	-	SCREEN 1,1
6	640 * 200 pels	16k	2	-	SCREEN 2

Notes:
> a) text modes also permitting character blinking
> b) character positions * screen lines
> c) 16 foreground colours, 8 background colours
> d) horizontal * vertical pixels

text and three of graphics. Each mode is identified with a mode code which in some cases may be provided by the operating system or by commands built in to high level languages (eg. BASIC). A summary of the modes is shown in Table 6.3. The CGA RAM has a start address of 0B800H, although different amounts of the memory are used for a screen image in different modes. Thus the text modes allow several display "pages" to be created, although only one may be on screen at a time. In BASIC the SCREEN command may be used to specify which page receives output (from PRINT commands) and which page is currently on screen. The CGA's character generator forms characters using an 8*8 dot matrix (irrespective of the number of characters per line), and this does not lead to particularly well formed characters. In the graphics modes one essentially trades resolution for colour, as the 640 * 200 pixels display is only available in black and white, while the multicolour displays can only be described as poor resolution. The CGA generates signals for displaying up to 16 colours using 200 raster lines at a rate of 15.75 kHz. The signals are produced as red, green, blue and intensity levels, together with horizontal and vertical synchronisation signals, and may be displayed on an RGBI colour display monitor.

The Enhanced Graphics Adaptor (EGA) can utilise all the display modes of the CGA and monochrome adaptors and four enhanced modes - although the latter are not accessible from all versions of BASIC. The enhanced modes are shown in Table 6.4 and use adaptor RAM starting at 0A0000H or 0A8000H. When generating enhanced mode signals the EGA produces up to 64 colours and either 200 raster lines at a rate of 15.75 kHz (suitable for a standard RGBI colour display monitor) or 350 raster lines at a rate of 21.8 kHz (which requires an enhanced colour display monitor). The signals are produced as primary red, green, blue, secondary red, green, blue, and horizontal and vertical synchronisation signals, and it is important to realise that an enhanced colour display or multisync monitor is required to utilise all these signals.

Table 6.4   Additional display modes with the EGA

mode	display	RAM	colours	QuickBASIC commands
13	320 * 200 pels	32k	16	SCREEN 7
14	640 * 200 pels	64k	16	SCREEN 8
15	640 * 350 pels	64k	mono	SCREEN 10
16	640 * 350 pels	128k	64	SCREEN 9

The Professional Graphics Adaptor (PGA) can generate 640 * 480 pels displays in up to 256 colours. Unfortunately the monitors which can adequately display images of this resolution are relatively expensive. While many drawing and circuit design packages support the PGA, relatively little general business software does. However, the PGA does provide all the facilities of the lower resolution displays.

It is important to realise that the display adaptors are controlled by the user's software, which selects the display mode and, in the case of graphics displays, sets the necessary bits. As the data required by the adaptor is different for the different modes, it is not surprising that much commercial software is able to use only a limited number of modes. At the present time most software (languages, word processors, editors, etc.) may be set up to work with the monochrome or Hercules adaptor and the CGA. An increasing amount of software (particularly packages from established companies such as Microsoft, and graphics software such as CAD programs) may also be set up to run with the EGA. However, having an EGA does not ensure that a particular program will be able to use it in one of the enhanced modes. Thus Microsoft's BASIC and CHART will use an EGA in CGA mode.

In addition to these "standard" display adaptors a number of non-standard adaptors are available, both from the manufacturers of PC clones and from manufacturers of special display systems. A number of these offer attractive enhancements over the standard display adaptors, including a greater variety of colours or an enhanced resolution on screen. The principal difficulty one encounters in using such adaptors is that some software will not be able to make use of the non-standard display enhancements, and may not function correctly even in the "standard" display mode. Thus the Genius full page display is a very high quality monochrome display system which can offer IBM monochrome mode, CGA simulation (in monochrome), and a full 66-line text and 728*1008 pixel graphics display for use with GEM, Windows, and a number of popular applications packages. Unfortunately a number of "badly behaved" CAD packages are unable to use the full screen display, and are either unable to work at all with the CGA emulation or have serious deficiencies - presumably because they attempt to access hardware features of the display adaptor which are not present.

6.3 Interface cards - a user port adaptor

The PC normally has no external connections other than for the power supply and the keyboard. Any connection required between the PC and an external device is provided on a

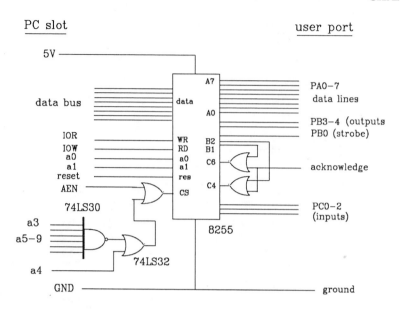

Fig 6.4   *An interface adaptor based on an 8255 PPI. This circuit allows the signals available on a PC's expansion bus to provide the eight data bits and two control lines required for the popular user port connections.*

card plugged into one of the expansion slots, usually in the form of a connector protruding through the PC's cabinet. Most PCs will contain one of the video adaptors described above, and many of these also provide one of the standard interface connectors described in chapters 8 and 9. A standard interface connection is usually provided so that the PC can communicate with a modem or output information to a printer or plotter, and in many cases additional cards may be used to provide extra standard interface connections (which may be useful if more than one printer interface type is required, or if both a printer and a plotter are to be used). However, most of these standard interface facilities are provided on the assumption that only one device will be connected to each interface port. For laboratory applications a more versatile interfacing port is desirable.

Many earlier micros were fitted with a bidirectional parallel handshaking interface known as a user port. While not one of the precisely defined standard interface systems, a user port is so useful for interfacing to laboratory systems that a number of adaptors which provide typical user port facilities are available for use with the PC. An example of a user port adaptor is shown schematically in fig 6.4, and dozens of this design are in use in our laboratories. A particular attraction of this design is that it may be operated directly from BASIC for use in slow applications or teaching, while fast machine code routines may be used for more demanding applications, such as fast (<10000 bytes s^{-1}) control of instrumentation. A second attraction of the system is that a simple user port may be used to communicate with several devices (as described in chapter 7), allowing a computer to operate quite complex instruments, and indeed several user ports may be fitted inside a single PC (even on a single card), so that one PC may be used to operate several instruments.

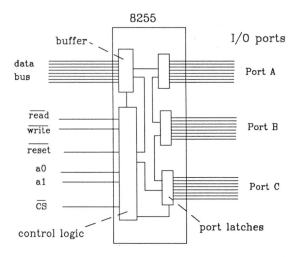

Fig 6.5   *A simplified view of the 8255 PPI, showing the registers which control the chips operation and provide the IO lines of ports A, B and C.*

Byte transfers through the user port may be accomplished in several ways, although we have adopted three techniques which cover most requirements for byte input, byte output, and bidirectional byte transfer. To understand the operation of the user port circuit in fig 6.4 it is necessary to digress for a description of one of the integrated circuits designed for use with the 8086 family of processors, namely the 40 pin 8255 programmable peripheral interface (PPI) shown in a highly simplified form in fig 6.5.

The 8255 may be regarded as consisting of four 8 bit registers, three of which (A-C) are connected directly to pins on the IC and so may be used as 8 bit wide inputs or outputs, and one of which (the control register) is used to control the operation of the other three by responding to bytes placed in it by the computer. The 8255 may be programmed (through this control byte) to operate in a large number of modes, although essentially these are combinations in which either the individual registers are used for input or output without handshaking, or registers A or B are used for input or output with handshaking - the handshaking signals being handled by individual bits in register C.

For full details of the operation of this versatile IC the reader is referred to Intel's data sheet for the 8255, but an indication of the steps involved in handling data transfers may be obtained from an examination of the techniques we use to program the 8255-based user port adaptor of fig 6.4. The 8255 is initialised (ie registers A, B and C are configured as inputs) by the application of a logic 0 level at the chip's reset pin (35). In fig 6.4 this pin is connected directly to the reset line of the PC's expansion bus, so that the 8255 is initialised every time the PC is reset (which occurs on power-up and when the operating system is restarted). The chip is accessed using five signal levels, one on the chip select pin (6), two on the address lines a0 and a1 (pins 9 and 8 respectively), and one on each of the chip's read and write lines. The chip select signal must be a 0 and either the read or write signal (but not both) must be a 0 for the 8255 to respond.

For a read operation (named from the PC's point of view) the 8255 copies a data byte from a register onto the data lines, while for a write operation a byte from the data lines is copied into a register. Which register is used depends on the 2 bit number decoded from the chips two address lines a0 and a1 (00 indicating register A, etc.). Which operation (ie read or write) is performed depends, not surprisingly, on whether the read or write line is a 0. This procedure is used both for sending control bytes to the 8255's control register and for data transfer between the PC and the chip's port registers.

In the user port adaptor the chip select signal is derived by "decoding" the expansion bus address lines a3 - a9, (a0 and a1 are used directly by the chip, and a2 is not used at all). The decoding is achieved by using NAND and OR gates to produce a 0 level whenever the required values are present on bits 3 - 9 of the address bus. In fig 6.4 the circuit produces a 0 whenever a4 is a 0 and a3, a5, a6, a7, a8 and a9 are 1s. Note that it makes no difference what values are contained in bits a0-a2 and a10-a19 of the address bus. Valid addresses are thus XXXXXXXXXX1111101XXX (where an X indicates irrelevant), and the lower half corresponds to address 3E8H-3EFH (1000-1007 decimal).

The decoded address is ORed with the address enable (AEN) line of the expansion bus, producing a 0 on the 8255's chip select pin only when the address bus contains a value within the required range and the system has signalled that the address is valid (which it does by holding AEN low). Once the 8255 is selected it decodes address lines a0 and a1 to determine which of its four registers will be involved in the operation. If IOW is low the 8255 proceeds to copy the byte on the data bus to the selected register, while if IOR is low the chip copies the contents of the selected register to the data bus. If neither IOR nor IOW is low then the chip does nothing - and this situation will arise whenever the processor (or anything else using the expansion bus) generates a read or write address within the range discussed above but intended for other circuits (such as RAM). If both IOR and IOW are low then something is wrong with the computer!

We adopted the user port approach of fig 6.4 because a) we wished to maintain hardware compatibility with a laboratory interface system developed for use with Commodore and BBC microcomputers, and b) we preferred to keep experimental electronic circuits outside the computer's cabinet - just in case. Figure 6.6 shows the layout of the user port adaptor used on a variety of PCs in our laboratory. On an in-house card the adaptors cost only a few pounds to produce and have the advantage of providing a highly effective interface port for instrument control (see chapter 7). In practice we have used 37-way D connectors on these cards, for the only reason that most PCs are already bristling with 25-way connectors for both serial and parallel ports. The presence of so many spare pins on the connector also provides a useful means of making additional connections when required (eg. to obtain power supplies or multiple user ports).

While describing the circuitry which allows the computer to access the 8255's registers has been long-winded, accomplishing these accesses in software is remarkably simple. To send a byte, n, to register A of our user port adaptor from a BASIC program we simply use the command

OUT 1000,n

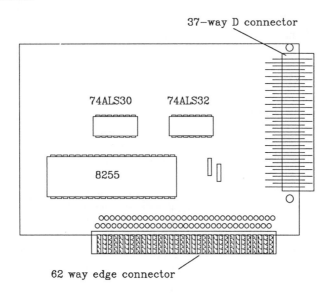

Fig 6.6 *The layout of a PC user port adaptor card.*

Similarly to read a byte from register C (and store it into variable X%) we use the function

$$X\% = \text{INP}(1002)$$

The analogous commands in assembler for transferring a byte to or from the lower half of the processor's ax register are

```
        mov     dx,1000
        out     dx,al
and
        mov     dx,1002
        in      al,dx
```

User ports make use of 8 data lines for byte transfer and at least two control lines, one to signal from the user port to an external device, and one to bring a signal from the external device to the user port. The user port adaptor of fig 6.4 uses the 8255's A register to store a byte which mirrors the levels of the user port data lines; the A register may be configured for input or output depending on our requirements at any particular time. For outgoing control lines we use the bits of the B register, with bit 0 being the mandatory signal line in fig 6.4. This carries the signal usually referred to as the "strobe" signal during an output byte transfer. Thus register B is nearly always configured to provide outputs only. Incoming control signals are handled by register C, and it is possible to configure register C so that all its bits are inputs. However, we rarely use register C in that manner, as it is convenient to make use of some of the more sophisticated modes of the 8255's operation so that particular bits of register C respond to transitions on incoming handshake lines. This approach is useful because many external device interfaces are designed to transmit a brief "acknowledge" pulse when they have correctly received and latched a data byte from the data lines, so it is sensible to be able to detect the pulse (ie the transitions between levels) rather than to rely on being able to observe the short duration change of a single bit which could be produced during the acknowledge pulse.

Reference to fig 6.4 shows that the A register pins are connected to the user port data lines and bit 0 of the B register is connected to the output control line labelled PB0. The control input labelled PCx is connected to both bits 4 and 6 of the C register - but through a pair of OR gates, so that the bit which actually receives the PCx signal may be determined by software. Either or both of bits 4 and 6 may be maintained as 1s by adjusting the levels being output from bits 1 and 2 of the B register; for example, a 1 being output from bit 1 of the B register maintains the level at bit 4 of the C register permanently at 1, irrespective of any changes on PCx. Bits 4 and 6 of the C register thus need to be configured as inputs at all times (because they are connected directly to logic gate outputs).

Writing a byte value of 137 to the 8255's control register configures all 8 bits of registers A and B as outputs and all 8 bits of register C as inputs (although not transition-sensitive inputs). This mode of operation may be used to send bytes from register A to an external device under the control of a single control line (PB0) without handshaking. Typically the user's program will output the data byte and then toggle PB0, which simply changes from a 1 to a 0 and back to a 1 again, to inform the external device to collect the data byte (which the external device must do on the 1 to 0 transition of PB0). A BASIC routine for this sequence is

```
100    OUT    1003,137         :REM configure 8255
110    OUT    1001,1           :REM ensure PB0 high
120    OUT    1000,data        :REM output data to register A
130    OUT    1001,0           :REM set PB0 low
140    OUT    1001,1           :REM return PB0 high
150    RETURN
```

Note that this arrangement ensures that a data byte is always present on the user port data lines - the last byte to be sent to the A register. Used in this way both the 8255 and the user's program ignore any activity which may be present on the PCx input line.

Sending the byte 153 to the 8255's control register configures registers A and C for input and register B for output. This configuration may be used for byte input (without handshaking) from external devices which will place a byte on the data lines on receipt of a 0 on a single control line (PB0). A simple BASIC routine for input of a byte is:

```
200    OUT    1003,153         :REM configure 8255
210    OUT    1001,1           :REM ensure PB0 high
220    OUT    1001,0           :REM set PB0 low to request data
230    X% = INP(1000)          :REM input byte and store in X%
240    OUT    1001,1           :REM return PB0 high
250    RETURN
```

Again any activity on the PCx control line is ignored.

To use handshaking for byte transfers we must configure the 8255 so that bits of the C register respond to transitions on the PCx control line. This may be achieved by sending a value of 193 to the control register and sending a value to register B to select which of the bits of the C register will receive the PCx signals. It should be noted that in this mode some bits of the C register are configured as inputs and some as outputs. In its handshaking operations the 8255 behaves in a more complex manner than in the simple input and output modes described above, so we need to consider the handshaking input and output operations separately. Also of course, the techniques described presuppose that the external device which is partaking in the handshaking operations does use the same handshaking technique. The hardware for ensuring this is discussed in chapter 7.

First we will consider user port byte output with handshaking, and for this the PCx control line must be connected to bit 6 of the C register - so bit 2 of the B register must be a 0 while bit 1 must be a 1. When a byte is written to register A, bit 7 of the C register is automatically set lo (ie changed from a 1 to a 0). [The pin connected to this bit may be used as a control line, although, as we prefer to have direct control over the signal levels produced by our circuits, our user port adaptor does not make use of this facility.] However, the byte value written to the A register does not actually appear at the user port data lines - the A register's output is disabled (it is a tri-state output buffer). In our circuit the user's program must now toggle the PB0 control line to indicate to the external device connected to the user port that it should receive some data. When the external device responds by lowering PC6, this enables register A's output (so the data now appears on the port data lines and can be read by the external device). When the external device returns PC6 to a 1, register A's output is again disabled and bit 7 of the C register is automatically set to a 1 (which indicates that the output buffer has been read by the external device). If the user's program writes another byte to the A register, then the 1 in bit 7 of the C register is automatically cleared to a 0.

A simple BASIC routine for output of a byte with handshaking is:

```
300   OUT   1003,193          :REM configure 8255
310   OUT   1001,3            :REM ensure PB0 high
320   OUT   1000,data         :REM output data to A register
330   OUT   1001,2            :REM set PB0 low
340   OUT   1001,3            :REM return PB0 high
350   IF (INP(1002)AND128)=0 THEN GOTO 350
360                           :REM wait for pulse on PC6
370   RETURN
```

Handshaking bytes from an external device into a user port may be accomplished in a number of ways, depending largely on whether it is the computer or the external device which is permitted to initiate the process. As part of our design philosophy we always adopt the approach that the computer (or rather, the user's program) should initiate activity for communication with laboratory instruments, and for that reason the handshaking on input technique described here is somewhat different from that normally associated with computer peripherals. An alternative approach is described in Intel's data sheet for the 8255. Our technique calls for the computer's user port to transmit a request for data signal on the PB0 line and then wait to receive an acknowledge pulse on the PCx line. This can be implemented by directing the PCx signal to bit 4 of the C register (which requires that the B register is outputting a 0 for bit 1 and a 1 for bit 2). In this configuration a low to high transition received at bit 4 of the C register latches the content of the user port data lines into the A register and sets bit 5 of the C register to a 1 (to indicate that data is available for reading). Reading the A register automatically clears bit 5 of the C register. A simple BASIC program for handshaking a byte into the user port is:

```
400    OUT    1003,193              :REM configure 8255
410    OUT    1001,5                :REM ensure PB0 high
420    OUT    1001,4                :REM set PB0 low
430    OUT    1001,5                :REM return PB0 high
440    IF (INP(1002)AND32)=0 THEN GOTO 440
450                                 :REM wait for pulse on PC4
460    X%=INP(1000)                 :REM read A register & store in X%
470    RETURN
```

Additional ways of using the user port adaptor and some of the signals which may be used on the remaining B and C register lines are discussed in subsequent chapters.

Finally it should be noted that the 8255, in common with most of the members of the 8086 family of chips and supporters, is produced in a variety of qualities and types. C-MOS versions (eg. the 82C55-8) are available for low power equipment, and I have used these for user port adaptors which occupy the modem space of Toshiba lap-top micros. But, and much more importantly, it is essential to ensure that 8255s used in circuits of the kind shown in fig 6.4 are matched to the speed of the processors which read and write to them. The standard 8255 will only work correctly with PCs operating at their original speed (4.77 MHz), while to operate with a modern AT an 8255-8 or 8255-10 is required to cope with the 8 or 10 MHz operation respectively.

6.4 The disk operating system

The PC is actually capable of running programs which have been prepared in a variety of languages and then compiled to produce machine code versions of the program. Many commercially available programs are supplied in this form on disk. In order that machine code programs can be used on a variety of different PCs, most of the operations which are dependent on the hardware arrangement of the PC (such as displaying characters on the screen, obtaining characters from the keyboard, and storing characters in disk files) are actually handled by a special control program called the operating system. Partly because the operating system is supplied on disk for loading into the PC each time it is switched on, and partly because the operating system handles the processes involved in keeping files on disks, the operating system most commonly encountered on a PC is called a Disk Operating System or DOS.

While there are several popular versions of DOS for the PC type computers, the version used on IBM PCs and that most widely used on other makes of PC is that written by

Microsoft. The version used on IBM PC is called PC-DOS, while the version sold by Microsoft to other computer manufacturers is called MS-DOS. For our purposes PC-DOS and MS-DOS may be regarded as the same operating system.

When a PC is turned on a short program inside the PC (held in ROM) is automatically started. This is called the BOOT program because it tries to pull in the operating system (by its bootstraps). The boot program looks for a disk drive (which must be present) and attempts to load the operating system (DOS) programs from a specific location on a disk in the drive. If there is no disk in the drive the boot program displays a message asking the user to insert a disk. If the disk in the drive does not contain the operating system program, then the boot program displays another message. If the disk does contain the DOS programs, then these are loaded into the computer's memory and run. When DOS is running the BOOT program can be restarted by pressing the <ctrl>, <alt> and <del> key simultaneously.

DOS is actually a collection of programs, some of which remain in the computer's memory until the power is turned off, while others (particularly the larger and less frequently used ones) are left on disk. DOS programs are run by typing a command on the keyboard. If a command is typed which refers to a DOS program in memory - known as an internal command - then the program is run immediately. However, if the command refers to a program which is not in memory - an external command - the DOS will attempt to load the necessary program from disk. Of course, if the disk containing the relevant command program is not in the drive, then DOS won't find it and will report an error (usually "bad command or file name")

A typical internal command is DIR (short for directory), which causes DOS to read the disk directory and display it on the screen. This is a fairly frequent requirement, so it is not surprising that the DIR program is kept in RAM. A typical external command is FORMAT (which formats a new disk - erasing any information previously stored on the disk). Formatting disks is not (normally) carried out as frequently as displaying a disk directory, so it is not surprising that the FORMAT program is kept on disk and only loaded into RAM when required. Internal DOS commands (ie those which do not appear in the directory listing) are actually stored in a file called COMMAND.COM - which is probably the first file in the directory. COMMAND.COM is actually copied into RAM when DOS is loaded, and this enables commands such as DIR and TYPE to be used even when the MS-DOS disk is replaced by another disk.

It is useful to bear in mind that anything typed at the keyboard immediately after the DOS prompt is interpreted by DOS as a command. DOS examines what has been typed to determine whether it is an internal command. If it is, then DOS attempts to carry out that command. If it isn't, then DOS examines a disk to see whether a file with the appropriate name is available for loading. Thus the DOS external commands and user programs are treated equally - as programs to be loaded from disk and executed.

6.5 The DOS interrupts

The DOS commands are those features of the operating system which every PC user becomes familiar with fairly quickly. However, it is the other aspect of the operating system which is in reality the most valuable and which has been responsible for the very high degree of compatibility between PCs manufactured around the world. In addition to providing utility programs which may be operated by the user from the keyboard, DOS also provides the programmer with an "interface" which allows him to use the hardware of the machine in a way which does not require a profound knowledge of that hardware (and of the many different forms in which it can appear). Thus DOS allows the programmer to display a character string on the screen without requiring any knowledge of the address of screen memory, the format of the screen or the kind of display adaptor connected to the machine his program will be used on. It allows him to store data in disk files without requiring him to know in advance whether the disk is hard or floppy, high density or low density, or 5.25 or 3.5 inch diameter. The programmer can pass a message to DOS requesting that a file is opened, a string is written to the screen, or a character obtained from the keyboard, and then rely on the operating system to know how to perform the task with the hardware available.

In practice communication with DOS is carried out by calling a software interrupt. In assembler this is achieved using the INT instruction, and in higher level languages by calling routines or functions defined for the specific language which themselves generate an INT instruction. When an INT instruction is performed the processor begins to carry out a specific routine of instructions pointed to by the contents of a specific address in memory - just as was the case for the hardware interrupts we discussed in chapter 5. In the case of software interrupts the interrupt routine transfers control to the DOS program in RAM and the DOS program discovers what the user's program required by examining the content of processor registers loaded by the user's program before the INT instruction was carried out. Most of the services provided by DOS are accessed by loading the processor's AH register with a function code, then issuing the instruction INT 21h. Thus if the AH register is loaded with 2 and INT 21h is obeyed, then DOS will attempt to display a character on the screen. The character it displays will be the one for which the ASCII code is found in the DL register; so the sequence:

```
mov     ah,2
move    dl,"?"
int     21h
```

will cause a ? to appear on the screen at the position previously occupied by the cursor (whether visible or not). In principle this should be true for any micro operating under MS-DOS and should be completely independent of the type of display system used on the micro.

If the last paragraph was cluttered with non-familiar terms - fear not. The argument is equally true for the BASIC command PRINT "?", although in this case there is always room for the suspicion that it is the BASIC system which actually arranges for the command to work on almost any computer. The important point is that the operating system provides a mechanism by which simple commands can be issued from a program to carry out tasks which in reality require a profound knowledge of the structure of the hardware on which the program is running. Or, to put it another way, the operating system makes the programmer's task much easier. DOS has built-in functions to handle disk files, program termination, the date and time, the keyboard, text on screen (unfortunately not graphics), and a number of standard interfaces provided on cards, such as the serial ports and parallel printer ports. Most such functions are available in many languages. For example, DOS function 44h passes control information from a user's program to a device driver program resident in RAM and may be called using INT 21h. Almost the same facilities are available in GW-BASIC by using the IOTCL command and function, even though that BASIC keyword is probably less familiar than, say, PRINT or INPUT.

While the advantage of writing programs which access parts of the computer through DOS (rather than attempting to access the hardware directly) is that the program has a much better chance of working in the same way on any other micro running DOS, there are some disadvantages. For example, DOS was written on the understanding that no more than two serial ports and three parallel ports would be in use on a single-user PC at any one time. If one wishes to have six serial ports or five parallel ports in use at once, then some other software must be responsible for handling their communication. Just asking BASIC to OPEN "COM1:...." to OPEN "COM6:....." will not work. Similarly if one uses an interface card which is not recognised by DOS as a standard device, then software which communicates directly with the card is required, either in the form of a "device driver" (accessed by IOCTL commands from BASIC) or specific subroutines which handle byte transfers to and from the interface. While many interface cards could be designed to work in place of standard cards, in practice most manufacturers and in-house constructors have accepted the relatively minor inconvenience of needing non-standard software to communicate with non-standard cards. Thus the user port adaptor described in section 6.3 is used by calling the short subroutines described for byte input and output, rather than OPENing as channel to the card and using INPUT#n and PRINT#n commands.

6.6 Other environments

The main restriction of DOS is that it is character based. It was designed to allow strings of characters to be received from the keyboard or serial port and sent to the screen and printer - and, of course, to handle transfers to and from disk files. As a character based system it is highly successful; after all one can issue PRINT#n, "HELLO" (or the equivalent in most other languages) to almost any display screen or printer and it generally works. However, much of the information displayed and output from laboratory systems consists of graphical data, and DOS was not designed to provide a hardware-independent approach to this. Two major attempts were made to provide a software environment in which a programmer could (metaphorically) say "draw a box", and the system would actually draw a box, on a CGA or EGA screen, on a matrix printer or on a Postscript printer. One attempt led to the GEM system by Digital Research, and the other to Microsoft's Windows.

From the user's viewpoint there are similarities between GEM and Microsoft Windows, although they are incompatible with one another. Both provide most of the facilities of an operating system, but with a more "friendly" user interface, showing lists of file names (or, in the case of GEM, pictures of "folders" with names underneath) from which the user can select by clicking a mouse or other pointing device. Both provide a consistent windows image (ie bordered areas of the screen which represent different disk directories or the display messages to the user), and both will operate on a very wide range of display devices and support a range of pointing devices and output devices (printers and plotters). Both may be used with a range of specific applications (including word processors which display different print styles, drawing and painting packages, and a variety of minor office-type applications), which are integrated and allow data from one application to be passed to another. Both will permit almost any program which will run under DOS to be initiated by clicking the pointing device at the directory display, and both will only provide device-independent graphics output for software which has been specially written to run in their particular environment.

For some years both of these rival systems have had supporters and opponents, and some major items of software have appeared which require one of these environments. [Thus, Ventura Desktop Publisher Edition runs under the GEM system, while Aldus Pagemaker, Micrografx In-a-Vision, and Hewlett Packard's Scanning Gallery operate in the Microsoft Windows environment.] With the announcement of Windows 2 and Windows 386 it would seem that Windows has a more secure future. In any event Windows is an important environment for IBM-compatible microcomputers and we will concentrate the remainder of this discussion on this system.

From the programmer's viewpoint, Windows is the environment in which a program operates. In principle the environment is very simple - although in practice it is .... well, usually described as having a steep learning curve. While most programs which run under DOS can be started from the main Windows display, programs which run under Windows must be designed with a somewhat different philosophy. Programmers familiar with object-oriented languages (eg Smalltalk) would probably regard the Windows environment as relatively straightforward. Those more accustomed to procedural languages probably think otherwise. Essentially a Windows program is a loop which looks for incoming messages (from a mouse or the keyboard) and takes action according to the nature of the message. The action may involve calculation, disk or interface activity, or the transmission of a message to the Windows system to request a modification to the display or the production of hard-copy output. The attraction of the technique is that it is not necessary for the user's program to know what type of keyboard or pointing device is generating the inputs, what type of display adaptor is actually displaying the output, or what model of printer is to generate the hard copy. This leaves the programmer free to concentrate on the activities of his own program without being over-burdened with a large number of alternative IO routines to cater for all possible IO devices.

While programs intended to run under Windows may be written in a language such as C, the interface with Windows is achieved through calls to the Windows routines. These addresses for these calls may be generated by linking the user's object module with a Windows library available in the Windows Development kit (available from Microsoft). In fact a program written in this way tends to consist of a very large number of calls to Windows routines, together with a relatively small number of calls to the user's own calculation and disk IO routines. Part of the art is getting to know which Windows calls to make; after all, there are several hundred routines in the latest development kit.

An attractive alternative approach to writing programs which run under Windows is to use a very high level language system know as Actor (produced by the Whitewater Group). Actor is available as an integrated development system, including editor, debugger, and interactive system. This allows users to write some code, test it, find the bugs and correct them, then move on to the next bit of code, while remaining within Actor. Furthermore, Actor is an object-oriented language, and newly defined objects essentially become part of the language, so that having written the code required to, say, generate a specific type of graphical display, or handle records in a database for one system, this code is available when the user comes to write another system. Actor is an impressively powerful and concise system (for example, to write a text editor similar to Microsoft's Notepad takes a single line of Actor code!) and is fully integrated into the Windows environment. Unfortunately the Actor development system is also a very large system, typically allowing only around 80k for user's programs in a 640k machine. However, versions of Actor for Windows 2 are able to make full use of the memory available to Windows and are not restricted to the 640k limit imposed by DOS and Windows 1.

CHAPTER 7

INTERFACING MICROCOMPUTERS WITH LABORATORY SIGNALS

We have seen that the measurement or control functions of laboratory instrument circuits produce or utilise signals which may be voltages, currents or pulses, while the microcomputer is a byte oriented device which produces or uses 8-bit parallel binary signals. To enable the signals of an instrument to form input or output data for a computer a third device is needed to translate one type of signal into the other. Such a device is called an interface and its role is illustrated in fig 7.1. An instrumental signal may be a constant value or may be time dependent, while the computer's signal must be one or more bytes specified at a particular instant of time. For this reason the translation carried out by the interface must be triggered in some way, either by the computer or by the laboratory instrument. In describing the basic types of laboratory interfaces (section 7.1), we shall assume that the computer initiates the translation process with a logic signal on a single control line (using, for example, the software control described in section 6.3). The subject will be considered in more detail in section 7.4.

7.1 Basic instrumental interface types

To accommodate the different types of instrumental signals discussed in chapter 2 (input, output, analog voltages, logic levels and pulses) a number of different types of interfaces are available. In this section we shall briefly consider the nature of each of the common types of interfaces in turn. Most are available commercially in a form suitable for direct connection to the more popular microcomputers, although, if high speed and high precision are not

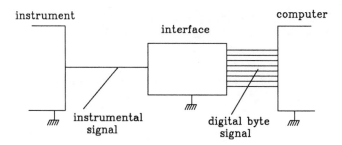

Fig 7.1 *The use of an interface unit to translate between laboratory signals and digital bytes.*

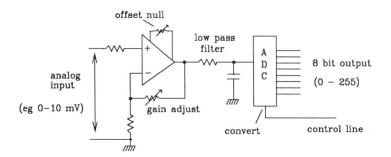

Fig 7.2 *A typical analog input interface.*

essential, interfaces can be constructed quite easily by a competent enthusiast or an
electronics technician.

7.1.1 Analog input interface

The analog input (or analog-to-digital conversion) interface translates constant or slowly
varying analog signals into bytes with binary values proportional to the analog input. A
typical analog input interface is illustrated in fig 7.2, where it will be seen that the circuit
consists of two stages. The first stage is an adjustable gain amplifier based on a non-inverting
configuration op-amp to allow a high input impedance. This converts the incoming signal
from the measurement range (say, 0 - 10 mV, the typical output range of a chromatographic
detector) to the standard input range (typically 0-1 V) of an integrated circuit ADC. When
the appropriate control signal level is applied to the ADC chip's "convert" pin, the analog
voltage at its input (0-1 V) is converted into a digital value (0-255) which is stored in an
internal buffer. The digital value can then appear as an 8-bit parallel digital signal on the
output lines, although this output is generally under the control of a signal applied to an
output disable pin or buffered through a TRI-STATE octal latch, so that the output can be
disabled when not required. By adjusting the gain and nulling the op-amp's offset voltage, an
8-bit analog-to-digital interface can be set to produce binary 0 output for 0 V input, binary
255 (ie. 11111111) output for 10 mV input, binary 127 for 5 mV input, and so on, (Note the
quantisation error).

Integrated circuit ADCs are normally voltage driven devices, so that if a current
measurement interface is required the voltage amplifier of fig 7.2 may be replaced by a
transresistance amplifier circuit (see section 3.5.5). Similarly if the peak value of an ac signal
is to be recorded the voltage amplifier may be replaced with a precision diode rectifier circuit,
provided that the output of this is adequately filtered to remove ripple before the signal is
applied to the ADC input. "True RMS converters" are also available in IC form for extracting
RMS value from a variety of ac input functions, although these are generally limited to
frequencies below 1 MHz.

The time taken by the ADC to convert the analog input into digital output, the conversion
time, varies widely with the type of ADC (and usually its cost). ADCs are available with
conversion times in the range of less than a microsecond to several milliseconds (see section

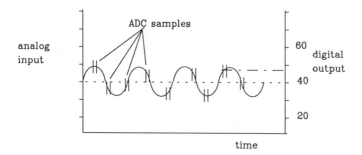

Fig 7.3 *An illustration of the effects of pick-up of mains-related interference by an analog input interface.*

4.7.2). Clearly the conversion speed required in a particular application will depend on the rapidity with which the analog signal is changing and on how often the signal must be sampled. Analog input interfaces are widely used for passing signals from the pen recorder output connections of instruments (such as chromatographs and spectrometers) to a computer, and in such cases a relatively slow ADC is usually adequate (eg. conversion time ca. 1 ms). Of course, unscreened leads which may have been adequate for connection to a chart recorder will need to be changed to properly screened cables and connectors for connection to an interface.

All ADCs are fast compared with, say, the frequency of the mains supply (50 or 60 Hz), so that mains frequency (or double mains frequency) interference and ripple on rectified signals can cause a problem. Figure 7.3 illustrates the problem; the interface under the control of the micro samples the analog input signal during a relatively short period of time, and the analog signal has a significant mains frequency interference component which results in widely fluctuating values of the digital input. The level of interference shown in fig 7.3 is not unusual on a signal intended for a chart recorder (indeed this author has purchased more than one brand of chart recorder which actually injected this amount of mains frequency ripple onto a 10 mV input signal). The problem can be usually overcome by filtering the analog signal with a simple RC network having a corner frequency of, say, 10 Hz - if such action will not mask the expected rate of signal variation. Alternatively, impedance conversion close to the signal source may be used to avoid interference pick-up on the analog input line.

7.1.2 Analog output interface

An analog output (or digital-to-analog conversion) interface performs the opposite translation, taking a byte value between 0 and 255 and converting it to an analog output signal. Two representative analog output interfaces are shown schematically in fig 7.4, where a) a voltage output DAC (such as the ZN428E) and b) a current output DAC (such as the AD7523) are illustrated. Again conversion from the binary value present on the 8 parallel data lines may be initiated by the application of an appropriate control signal level to the convert pin of the DAC IC, although not all DAC circuits provide this facility, and many offer a latchable buffer to hold the digital data so that an analog output may be maintained after a digital input signal has been removed.

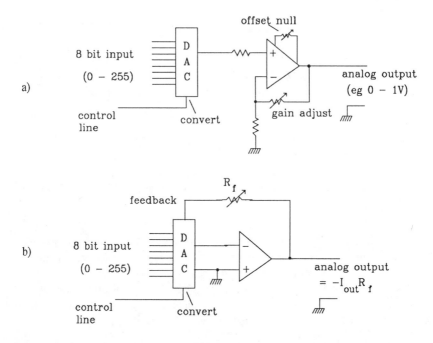

Fig 7.4 *Typical analog output interfaces. a) shows a circuit based on a voltage output DAC, while in b) a current output DAC is followed by a transresistance amplifier to generate a voltage output.*

Voltage output DAC circuits produce an output voltage typically in the range 0-1 V to 0-10 V, although most have relatively high impedance outputs which, to avoid temperature fluctuations, require buffering if a high accuracy output is required. The output voltage may be scaled by a selectable gain amplifier or attenuator to lie in any desirable voltage range (as illustrated in fig 7.4a), or may be passed to a transconductance amplifier to generate a required output current. Multiplying DACs allow a wider choice of output ranges, and the reference signal (which determines the output range) could be supplied by a second analog output circuit, allowing computer control of the range. Current output DACs produce an output which is typically in the range 0-1 mA, and which may be converted to an output voltage using a transresistance amplifier. Many current output DACs provide an on-chip feedback resistance for use with a transresistance amplifier, so that only a good quality op-amp is necessary to complete a voltage output circuit. An additional and variable feedback resistance may be incorporated (as in fig 7.4b) if a variable conversion gain is required.

DAC ICs are usually faster than comparably priced ADC chips and for most laboratory applications the speed of conversion does not form a limitation (see section 4.7.1). Analog output interfaces are frequently used to provide voltage signals for pen recorders and control signals for variable power devices such as lamps or heaters. They can of course be used to provide on/off signal levels, although generally the switched output interface would be more appropriate for this task.

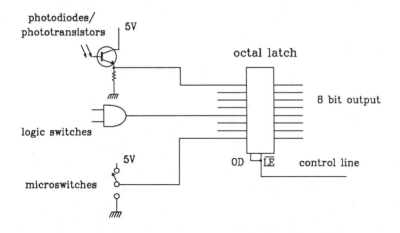

Fig 7.5 *A logic level input interface used to sense the input levels of several separate lines.*

7.1.3 Logic input interface

The logic input interface illustrated in fig 7.5 uses an octal latch to detect logic levels on up to eight sense lines. The signals in this case normally arise from relays, microswitches or logic circuits, although in some cases it is possible to use variable analog signal sources - provided that one remembers that the computer will only sense whether the voltage level is hi or lo (ie >2.5 V or <2.5 V for a CMOS latch operating at 5 V. TTL is not a good choice in this application unless there is no doubt about the ability of the source to sink the TTL lo level current). However, if there is a risk that a voltage level being sensed may lie in an undefined region (eg at 2.5 V for a CMOS system or in the range 0.8 - 2 V for TTL) then it would be better to convert the voltage to a logic level using a comparator with an appropriate comparison (threshold) voltage (as described in section 4.6.1).

In operation the hi/lo level of each of the eight sense lines is stored in the buffer as one bit of an 8-bit byte when the appropriate control signal level is applied to the latch enable pin of the IC. The byte is then read by the computer as a binary number, such as 10010111, where a 1 indicates sense level hi and a 0 indicates sense level lo. Thus a value of 255 would indicate that all sense lines were hi. Once the byte has been read by the computer the individual bit values may be decoded by software using the techniques described in section 5.1. The byte value represents the states of the sense lines at the time the latch enable signal is received by the buffer, a time which may be selected with considerable precision (ie. a specific microsecond) but may be earlier than when byte is read by the computer - particularly if the operation of the interface is being controlled from a BASIC program (see section 5.7). Typical applications are in safety devices (eg. ensuring that doors are closed before X-ray tube power supplies are activated, cf fig 4.4), sensing the presence or absence of a sample on a conveyor belt, and reading a manually switched control panel.

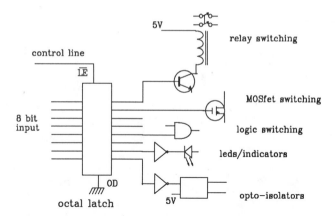

Fig 7.6 *A logic level output interface used to produce several logic signals suitable for driving relays, gates and switches.*

7.1.4 Logic output interface

A complementary type of interface is the logic level output or switched output interface (fig 7.6) which produces eight output logic levels of 0s and 1s, ie. nominal voltages of 0 and 5 V (of course, these may be readily converted to other voltages or currents). In this case a byte from the computer is stored in the octal latch when its latch enable pin is activated, and each line of the buffer output produces either 0 V if the corresponding bit is a 0, or 5 V if the corresponding bit is a 1. The latch used may be CMOS or TTL and, although the latch shown in fig 7.6 is a TRI-STATE variety (eg. 74C373 or 74LS373) with a grounded output disable pin, there is no reason why a non-TRI-STATE device should not be used. If only momentary logic level outputs are required, the output disable pin may be connected to the latch enable pin - in which case outputs will only be present when the control line is lo. This technique is useful when one wishes to operate an increase/leave unchanged type of control.

While these output signals may be used directly (eg. for operating low current indicators such as LEDs) it is usually better to buffer the output signals with transistors, TTL drivers or analog buffers such as voltage followers, if the load is likely to exceed a few mA, or to use logic activated, solid state relays or opto-isolator switches if it is necessary to switch high currents or voltages or ac. Opto-isolator switches are switches whose operation is controlled by the light produced by an LED housed in the same package as the switch. Opto-isolator switch ICs are particularly valuable for interfacing as they allow reliable electrical isolation between units attached to the computer and units attached to other powered systems. Typically potential differences of thousands of volts between the two circuits can be tolerated. In addition to simple switching applications, modern opto-isolator devices are available containing light-operated high gain transistors (Darlingtons), silicon controlled rectifiers (SCRs) and triacs. The latter are ideal for providing isolated logic level to mains interfaces. Switched outputs find application in automatic sampling and injection, in operating indicators and warning signals, in turning on and off controlled elements such as heaters or light sources, and in control of multiplexors and selectable gain amplifiers (see below).

7.1.5 Digital data interfaces

Both the logic input and logic output interfaces can, of course, be operated as 8-bit digital transmitters, and in this guise find application for communication with instruments which have facilities for 8-bit data transfer. These interfaces are often used with opto-isolator buffers, to minimise any risk of a fault in one system causing damage in another. The use of opto-isolator buffers also enables non-TTL logic levels to be interfaced to a computer with minimal difficulty. Many common laboratory instruments which have digital displays (eg. pH meters, frequency counters and digital multimeters) also provide multidigit BCD outputs. Typically these may take the form of 3.5 digits (ie. 0-1999) and a sign "bit" formed into 14 parallel digital outputs, along with a hi/lo "data valid" signal. This type of data can be transferred using a 16 bit digital input interface, such as that described in section 7.3 and fig 7.14. Of course if BCD or character data is transferred to a computer, then the user's program must be able to use or decode the bytes as required; thus the same byte values read from a BCD source and from a binary source represent different quantities (see section 5.1). 8-bit digital output interfaces are also useful when several different interfaces are connected to the same computer, and this subject will be discussed further in section 7.2.

7.1.6 Pulse counter interface

A pulse counter interface accepts pulses (usually logic pulses or pulses with specified characteristics which are converted into logic pulses) and counts them in a binary counter (see section 4.5.1). A typical pulse counter interface is shown in fig 7.7. The input pulses are converted into fixed width logic pulses using a 4528B monostable, in this case triggered by the positive-going edge of the input pulse. The width of the output pulse is chosen to suit the counter circuit, and in this case could be set to 1 microsecond by the choice of R=10 k and C=22 pF. The counter is an 8 bit binary ripple counter (4024B), which can only count from 0 to 255, and it is important to realise that pulse number 256 returns the byte value within the counter to zero, so that the user's program needs to check for this "byte overflow" and correct the values in the program accordingly (typically by adding 256 to a reading which is smaller than the previous reading). Furthermore the byte stored in the buffer represents the count at the time the latch enable signal is received by the buffer and will not change if further pulses are counted before the byte is transferred to the computer.

Pulse counter interfaces are of particular value in systems involving X-rays, radioactivity, photon counting or precision timing, although 16 or 24 bit interfaces are likely to be of more value than 8-bits when high pulse rates are involved (see section 7.3). Pulse rates or frequency may be determined using pulse counters by, not surprisingly, determining the count within a specified period of time. Time intervals can be recorded by using a pulse counter to count the pulses from a crystal oscillator (see section 7.4), and this is sometimes a more reliable technique than trying to use the microcomputer's own BASIC "clock" (if it has one), particularly on micros where the processor is interrupted frequently (many micros are interrupted 50 or 60 times a second). Of course, the MPU's clocking signal may be used as a source of pulses, but there is no reason to suppose that this will provide a high stability source of exactly the expected frequency. Furthermore, the signal used should be the control bus version, rather than the MPU's clock input, as most post-1976 MPUs have their clock

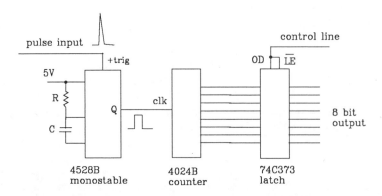

Fig 7.7 *A typical pulse counting interface, which may also be used for timing applications by counting pulses from a clock oscillator.*

circuits on-chip (only the crystal is externally connected). Some micros cannot tolerate capacitive loading of the clocking signal, so that the line requires buffering within the micro and before lengths of external cabling are attached.

7.1.7 Pulse output interface

Pulse output interfaces can also be constructed, although where pulses are used in instrumentation (eg. for stepping motors or clocked logic control) the height, width and rise time of each pulse is likely to be important, so that the interface in this instance is often supplied as an integral part of an instrument or sold for the control of a specific device (such as the Bentham stepping motor interface). Furthermore, two distinct categories of pulse output interfaces can be distinguished: those for which the output consists of a continuous stream of pulses or a square wave at a selected pulse rate; and those for which the output consists of a selected number of pulses. The former variety is useful where a particular frequency signal needs to be applied to a sample, or where a "chopping" signal is required. The latter category has applications in dispensing discrete samples of material and in providing pulse trains for the operation of stepping motors.

Typical circuits representing each type of pulse output interface are shown in fig 7.8. Both are based on CMOS 4526B programmable divide-by-N counters driven by a 1 MHz crystal oscillator, and both may have their outputs buffered to provide TTL-compatible pulses. The 4526B is a 4 bit device, but is designed so that several devices may be cascaded. Its "0" output (that is what it's called) is normally 0, but can change to a 1 when the count reaches 0000 (ie. when it overflows) if the level on the cascade feedback (CF) pin is a 1. Thus in fig 7.8a the high order nybble device has its CF pin connected to the 5 V supply, allowing the "0" output to change from 0 to 1 when the high order count reaches 0000 (it starts at the value set by b4-b7). This output is connected to the CF pin of the low order nybble device, so the "0" output of the low order device is unable to change until after the high order count has reached 0000 and the low order device subsequently overflows. The preset enable (PE) pin of each device is connected to the low order nybble's "0" output, so each time this goes high (which occurs each time the number of clock pulses specified by the

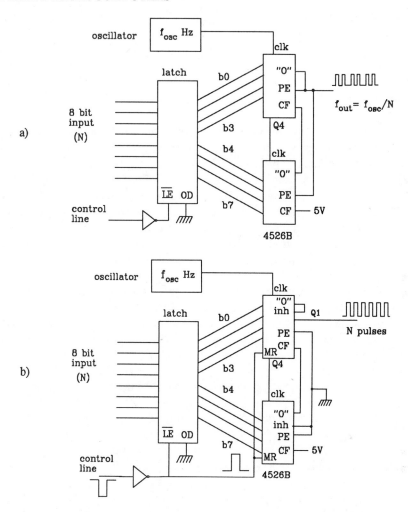

Fig 7.8 *Two examples of pulse output interfaces. a) generates logic pulses at a specified frequency, b) generates a specified number of pulses.*

data byte have been counted), the two devices reload their respective nybbles from b0-b7 and start counting again (which returns the circuit's output to 0). So the circuit in fig 7.8a produces a continuous stream of output pulses at an average frequency equal to the clock frequency divided by N, the value of the data byte. (Note that this technique does not produce equally spaced pulses for all values of N. If equal spacing is required it can usually be achieved by dividing down the output pulse rate using, say, a 4040 counter.)

In fig 7.8b the circuit's output is taken from the Q1 output of the low order counter, and this output changes level each time a clock pulse is counted - until that is, the preset count is reached and the device's "0" output goes hi and inhibits all the Q outputs by holding the inhibit pin hi. This can only occur after the high order device has counted its preset count, produced a hi output on its "0" output, and taken the low order device's CF pin hi. Thus the circuit produces the number of clock pulses preset by the data byte, and then does nothing until reset by a logic 1 on the master reset (MR) pin. Return of the MR pin to 0 allows another pulse train output of the number of pulses specified by the data byte.

7.2 Multiplexing

Each of the interfaces described above may be connected directly to an 8 bit port of a micro when a single control line is available. Thus unless the micro is fitted with several ports (as could be the Apple or the PC, through their I/O slots), we are limited to a single interface between the micro and any laboratory instrumentation. For most purposes this would be unsatisfactory. However we can overcome this limitation by using a computer's single parallel port as a bus, ie. connecting several different 8 bit circuits in parallel with the port. Naturally this requires a certain amount of care, as we must ensure that no more than one circuit attempts to set the digital signal levels on this bus - so that our circuit devices which can determine these signal levels must have TRI-STATE outputs.

One of the simplest techniques for the implementation of a bus can be seen in the part circuit shown in fig 7.9 (we will come to the missing items later). First of all we note that, in addition to the data lines of the computer port, three octal latches are connected to the bus, one latch (A) is connected at its input lines and the other two at their outputs. The latch enable and output disable pins of latch A are both connected to the computer port's control line, and we shall assume that initially this line is hi. The latch enable and output disable pins of the other latches are held hi by pull-up resistors (so the outputs of these latches are normally held in the high impedance state). Thus the computer is able to output a byte onto the bus without either B or C attempting to do the same.

Let us imagine that the latches B and C are holding 8 bit data which we wish to read into the computer via the bus. A typical sequence for carrying out the data transfers is as follows:

1. The computer outputs a byte with a binary value, eg. 254 (ie. 11111110).

2. The computer forces its control line lo to operate the latch enable and output disable functions of latch A. This results in 254 being latched into A and output appearing on the address output lines which carry individual bit levels to the output disable pins of the other latches.

3. The line from latch A (d0) connected to the output disable pin of latch B now carries a 0 (ie. lo level), so this latch latches the state of its input lines and outputs the byte value onto the bus. The other latches have their output disable pins held hi (d1 - d7 of latch A), so their outputs remain in the high impedance state. (There is actually bus contention at this point - both the computer and latch B are trying to hold bytes on the bus. For this technique to work the computer user port needs to be buffered with fairly tolerant bus drivers, while the latches in fig 7.9 should be CMOS. Under these conditions the computer wins without any damage being done, although the clash is short-lived anyway. We have used this technique for several years with PET, BBC

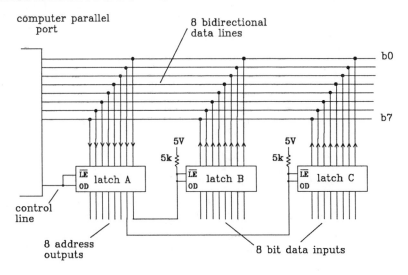

Fig 7.9 *A part circuit for addressing one of several interfaces connected to a bus. This system does result in bus contention, so the latches should be CMOS (eg. 74C373). Ways of avoiding the contention are shown in fig 7.12.*

and PC computers (using the user port adaptor) and experienced no trouble, but it could damage a computer fitted with feeble port drivers. An alternative arrangement which avoids this difficulty altogether is discussed below.)

4. The computer stops the output of 254 and reconfigures its parallel port for input. The data on the bus is now the byte value being output from latch B.

5. The computer inputs the byte value on the bus lines (eg. X = INP(1000) in the case of the PC user port adaptor of section 6.3).

6. The computer sets the control line hi, disabling the output of latch A, so that this latch is no longer holding the output disable lines of any latches. The outputs of the other latches become or remain disabled by the effects of the pull-up resistors connected to their output disable pins, so latch B is no longer asserting a byte value on the bus and the bus is free again.

Repeating the above sequence with a computer output of 253 (ie. 11111101) would allow the byte held in latch C to be read over the bus. As there are eight output lines from latch A we could easily arrange to read the contents of eight latches using this technique, each latch being addressed with a byte consisting of one 0 and seven 1s (corresponding to values of 254, 253, 251, 247, 239, 223, 191 and 127).

A somewhat more efficient way of decoding the addresses for latches connected to a bus can be based on a decoder IC. This approach is illustrated in fig 7.10, where the decoder is the CMOS 4515B device called a 1-of-16 decoder or demultiplexor. This particular IC utilises a

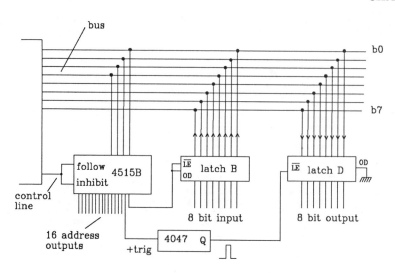

Fig 7.10   *Using a 4515B demultiplexor to address 1 of 16 single byte latches connected to a bus. Only the lower 4 bits of the bus are needed for addressing via the 4515B.*

4 bit input and in the circuit diagram the input pins are connected to the low order lines (ie. b0 - b3) of the bus. The IC has 16 output pins, S0 - S15, and normally 15 of these are hi and 1 is lo, the lo one being selected by the binary value of the 4 bit (0 - 15) input. Thus an input of 0 (0000) selects the output on pin 11 (S0) to be lo, and the other 15 outputs to be hi; an input of 5 (0101) selects a lo output on pin 6 (S5), and an input of 15 (1111) selects a lo output on pin 15 (S15). The device is not a TRI-STATE device, so it does not have an output disable facility of the type present in the octal latch we used above. However, the 4515B does have an output inhibit pin, and when the level applied to this pin is hi all the output pins are forced hi (level, not high impedance) irrespective of the digital input. Because the output is always hi when it is not specifically selected to be lo, the pull-up resistors used in fig 7.9 are not required when this decoder is used. There is also a "follow" pin (analogous to the latch enable connection of the octal latch). When the follow pin is hi the value decoded follows the coding of the 4 bit input, but when follow is made lo the value of the 4 bit code present at the digital input is stored internally and used for decoding until such time as follow returns hi again. Connecting the follow and inhibit pins together allows the circuit to be controlled by a single hi/lo control level from the computer.

If one considers the sequence of steps described for fig 7.9 while examining the interconnections of fig 7.10, then it will be seen that the circuits achieve the same results - but in the case of fig 7.10 we may select the contents of up to 16 different latches for transfer over the bus. Larger numbers of addresses can be used, for example, by using a second decoder with input taken from b4 - b7 of the bus, although in its simplest implementation this would require that one address for each decoder (such as 0000) was left unused so that the two decoders did not enable two different latches at the same time. Both of these address selection techniques produce the same end result and there are numerous other devices which could be used to perform this function. In many of the subsequent figures we shall use selectors based on inhibitable decoders (eg. 4515B) and TRI-STATE latches (eg. 74C373) more or less interchangeably. Of course when a TRI-STATE device is used in this particular role pull-up resistors are needed, although to simplify circuit diagrams these have been omitted in the remainder of this chapter.

An output latch D has been included in the circuit of fig 7.10 to illustrate one way in which an output can be obtained continuously (eg. for a chart recorder), while its 8 bit value may be changed at will. The output disable pin of this latch is grounded, so the device is always exerting an 8 bit output. However, the latch enable pin is connected to the output, Q, of a 4047 monostable, and as this level is normally lo the data present at the input pins of the latch is not normally passed through to the outputs - any previously latched data byte forms the output under these conditions. When the relevant decoder output (line 1 in this example) is selected it goes lo, and the required output byte may be loaded onto the data bus by the computer, although at this stage the outputs of latch D are not changed. When the decoder outputs are inhibited line 1 goes hi again. This lo-to-hi transition triggers the monostable which produces a lo-hi-lo pulse on its output, and this allows latch D to latch the data currently present at its inputs. As the OD line is still grounded the 8 bit output changes immediately to the newly latched value, and stays like that until the latch is enabled again by the decoder/monostable combination.

A typical BASIC subroutine to output a byte X through latch D via the user port adaptor described in chapter 6 is as follows:

```
100   OUT   1003,137:    REM configure port A for output
110   OUT   1001,1:      REM set control line hi
120   OUT   1000,1:      REM output address 1
130   OUT   1001,0:      REM set control lo
140   OUT   1001,1:      REM return control hi
150   OUT   1000,X:      REM output byte X
160   OUT   1001,0:      REM set control lo
170   OUT   1001,1:      REM return control hi
```

Similarly a subroutine for byte input is:

```
200   OUT   1003,137:    REM configure port A for output
210   OUT   1001,1:      REM set control line hi
220   OUT   1000,0:      REM output address 0
230   OUT   1001,0:      REM control lo
240   OUT   1003,153:    REM configure port A for input
250   X=INP(1000):       REM read port data bus
260   OUT   1001,1:      REM return control hi
```

The instructions used for control line toggling are described in section 6.3. A technique capable of an even wider range of addressing is illustrated in fig 7.11. One decoder connected to the high lines of the bus (b4 - b7) is used to select one of several (up to 16) decoders, each of which selects a latch from 1 of 16 using a code from the low order lines of the bus (b0 -

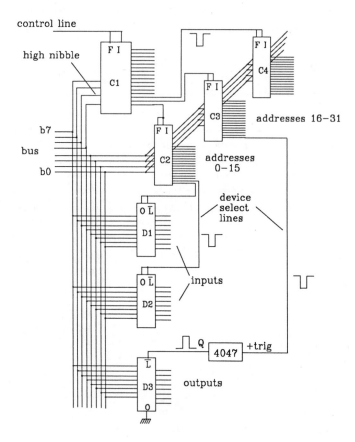

Fig 7.11 *An interface addressing system capable of extension to 256 addressed lines.*
The C_i are 4515B 1-of-16 decoders, while the D_i may be 74C373 latches or
any other latchable 8 bit input or output devices.

b3). This allows up to 256 different digital signals to be selected, which should be enough for even complex systems.

All of the above circuits produce a momentary clash on the data lines when any input interface is addressed. While as stated above this technique works perfectly well for PIA/PPI ports when CMOS TRI-STATE latches are used for the bus connections, there is a straightforward way of avoiding the clash - although at the cost of an additional control line operation and a consequent reduction in the speed of operation. The technique is illustrated by the part circuit shown in fig 7.12a. The control line is connected to a 4013B flip-flop (a circuit which can be arranged to change its output level for every hi-to-lo transition on its clock input), and to two OR gates. The control line is toggled hi-lo-hi with the required device address present on the data lines, and this results in a latch enable pulse being passed to latch A and the device address being latched. However, the outputs of latch A remain disabled until a second hi-lo transition occurs on the control line, and this is carried out only after the PIA port has been reconfigured for input. Once this occurs the addressed latch (eg.

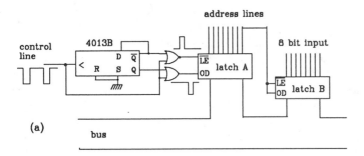

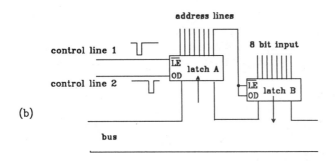

Fig 7.12 *Two ways of avoiding bus contention with an multiplexed interface system.*
a) The use of two togglings of a single control line to operate the address latch, A,
by alternate control of its latch enable and output disable lines. b) The use of two
independent control lines to operate the latch and output disable lines of latch A.

B) deposits its data onto the data lines where they can be read by the computer, and the PIA
control line returned hi to terminate the transaction. A typical BASIC subroutine for the input
of a single byte is as follows:

```
200   OUT   1003,137:   REM configure port A for output
210   OUT   1001,1:     REM set control line hi
220   OUT   1000,0:     REM output address 0
230   OUT   1001,0:     REM set control line lo
210   OUT   1001,1:     REM return control line hi
240   OUT   1003,153:   REM configure port A for input
210   OUT   1001,0:     REM set control line lo
250   X=INP(1000):      REM read port data bus
260   OUT   1001,1:     REM return control line hi
```

The corresponding control level changes are included in fig 7.12a. A similar arrangement can
cater for byte output without the use of the monostable pulse circuits of figs 7.10 and 11. And
of course the technique can be adapted for use with address decoders rather than address
latches by appropriate use of the follow and inhibit pins.

An alternative technique may be implemented by operating the address latch's latch enable and output disable pins with separate control signals from the computer as illustrated in fig 7.12b. This is a simple way of avoiding bus contention, but, of course, is limited to those micros on which a second output control line may be easily used. In the case of the PC user port adaptor a second control line is available (eg. PB1), although for many micros it is not so easy to find a second line. To maintain the compatibility of interface systems with a wide range of micros we have based most of our instrumentation interfaces on the single control line technique.

7.3 Multiple byte interfaces

An 8-bit interface uses data bytes with values between 0 and 255, and the smallest possible difference between byte values is 1 (eg. 122 and 123). An important consequence of this is that the resolution of the interface is only 1 in 255, equivalent to about 0.5% of the full scale value. While this may be acceptable in some instances (eg. counting infrequent events), it is easy to see that there are many cases in which the resolution of an 8-bit interface is inadequate. Figure 7.13 shows a chart record of an analog signal from a chromatograph, with the byte values produced by an 8-bit ADC marked on the signal scale. While the larger peak covers most of the 0-255 scale, and is therefore converted with reasonable precision, the smaller peak would be converted within the range 20-30 so that the peak height estimated from byte values would contain an uncertainty of around 10%. Probably the most common solution to this problem is to use a higher resolution interface, ie. an interface which converts the instrumental signal into 12, 16 or 24 bits. A 16-bit resolution ADC interface converts an analog input signal into a 16-bit binary number (with a value between 0 and 65535). As the computer can accept only one 8-bit byte at a time obviously the 16-bit data must be transferred to the computer as two separate bytes. This can be accomplished by storing the 16-bit data in two 8-bit buffers whose outputs are connected to a parallel bus as shown in fig 7.14. One buffer now deals with bits 0-7 of the 16-bit data, and the other with bits 8-15. This arrangement allows us to transfer bytes via the bus in much the same way as we did with the multiplexed 8 bit latches in section 7.2. Some BASIC instructions to transfer the two bytes and combine them to form the 16-bit value are shown below:

```
200    OUT    1003,137:    REM configure port A for output
210    OUT    1001,1:      REM set control line hi
220    OUT    1000,0:      REM output address 0
230    OUT    1001,0: OUT 1001,1:  REM control line lo then hi
240    OUT    1003,153:    REM configure port A for input
250    OUT    1001,0:      REM set control line lo
260    X=INP(1000):        REM read port data bus
270    OUT    1001,1:      REM return control line hi
280    OUT    1003,137:    REM configure port A for output
290    OUT    1001,1:      REM set control line hi
300    OUT    1000,1:      REM output address 1
310    OUT    1001,0: OUT 1001,1:  REM control line lo then hi
320    OUT    1003,153:    REM configure port A for input
330    OUT    1001,0:      REM set control line lo
```

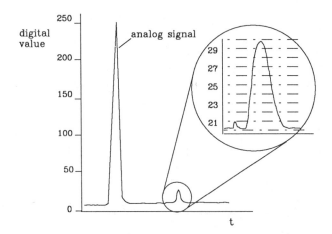

Fig 7.13   *The limitation of a low resolution analog interface is illustrated by the conversion of the smaller analog signal peak within a small digital range.*

```
340    Y=INP(1000):        REM read port data bus
350    OUT    1001,1:      REM return control line hi
360    Z = X*256 + Y:      REM Z contains 16 bit value
```

Note that the values stored in both X and Y must be in the range 0-255, while the 16-bit value produced in line 360 and stored in Z may be 0-65535. Most of the types of interfaces described earlier are readily available in 12 or 16 bit versions, and 24 or 32 bit versions can be obtained or constructed. For analog conversion interfaces 12 bits allows a resolution of 1 in 4095, 16 bits 1 in 65535, and 24 bits about 1 in 16 million. However, increasing the resolution of ADC and DAC interfaces does dramatically increase the cost of the interface, particularly if the retention of high speed conversion is necessary. The increased number of program steps required for multibyte data transfers also results in the speed limitations of BASIC soon becoming apparent. While up to 10 single data byte transfers per second may be achieved using BASIC with simple 8-bit interfaces in the manner illustrated above, it becomes difficult to reach 2 readings a second when 24-bit interfaces are used. So, unless the interface supplier also provides machine code software for data transfers, the user may have to write his own, or use a BASIC compiler, in order to make his program run at a useful speed.

7.4 Interface control

Interfaces with resolution greater than 8 bits and interfaces which are required to provide several quantities recorded at the same time, require more precise control of their functions than can be achieved with the circuits we have considered so far. For example, if data is to be read into a computer from a 12 bit ADC interface, then the 12 bit data must not change between the transfer of the first 8 and last 4 bits. One way in which this can be accomplished is illustrated in fig 7.15, where the ADC's "start conversion" signal is supplied by a monostable (generating, say, a 1 microsecond pulse, although the timing components are not

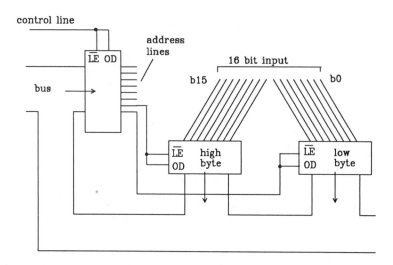

Fig 7.14 *A 16 bit digital input interface, which transfers its data over the 8 bit bus as two separate bytes controlled by two address lines.*

shown in the figure) triggered by the same select line as that which enables the output of latch C. The contents of both latches are latched by the ADC's "conversion complete" signal, which is assumed to be a lo output or a lo-hi-lo pulse (obviously an inverter could be included if the end of conversion was signalled by a complementary signal). Thus 8 bits may be read from latch C, then latch B selected and four bits read from this without the ADC's 12 bit data changing. For this circuit it would be necessary to ensure that the time delay between selection of latch C and the reading of the byte by the computer was longer than the conversion time of the ADC - not usually a problem when the controlling program is written in BASIC. Of course, it is also essential that latch C is addressed before latch B!

A more versatile approach involves dedicating lines from the address decoder for each function required by the interface. A fairly detailed example of this technique is included, both to illustrate the high degree of control which may be achieved, and to enable us to highlight a few aspects of timing accuracy. The circuit in fig 7.16 is for a multifunction interface consisting of two 24 bit counters, one which counts pulses from a measuring transducer (such as a photomultiplier - signal conversion circuits are not shown), and the other which counts pulses derived from a 10 MHz crystal oscillator, but divided down by 100 (a variable divider was used in the original to increase the useable timing range). This unit forms part of a system designed for the measurement of photon count rates (proportional to detected light intensity), and the clock was included so that high precision measurements could be made.

Let us consider first the decoder, which in fig 7.16 is a 4515B 1-of-16 decoder. Eight of its output lines are used (the original circuit had several additional functions), 6 providing latch select signals (a1 to a6) to enable the latches to deposit their data bytes on the bus, one providing a latch signal and one providing a reset signal. The latch signal was required to

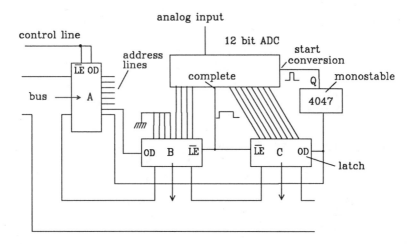

Fig 7.15   *Using an interface address to control the action of an interface system. In this*
case the ADC's "start conversion" line is pulsed (via the 4047) when the low
byte is addressed, and both latches are latched by the ADC's "conversion
complete" signal. The ADC used is the successive approximation ADC1210;
it requires additional circuitry (a clock) not shown.

operate all six latches at once, enabling each to latch the appropriate byte from the relevant
counter. Similarly the reset signal was required to reset all four (12 bit) counters to zero, so
that both the photon counter and the timer were simultaneously restarted from zero.

The latch and reset signal could be used directly by connecting these conductors to the
appropriate pins on the latches and counters. However, this would result in the time intervals
recorded containing an uncertainty of up to two oscillation periods (one for the reset at the
begining of the time period, and one for the latch at the end). This would have been
acceptable for timing periods of the order of 2^{20} cycles (ie. a few hundred seconds) but was
unacceptable for periods of less than a second (2^{10} cycles), for which the error was about
0.1%. A convenient way of improving the timing precision may be implemented by applying
latch and reset pulses to the latches and counters only on the falling edge of the clock's
waveform, which can be done by using the computer generated signals to enable a JK flip-
flop, whose output only changes when the edge of the clock pulse arrives at its clock input
(see section 4.6.2). The output's falling edge is then used to trigger the operation of a
monostable which produces an output pulse of 1 microsecond duration, which in turn is gated
by one of the two NAND gates and applied to either the latch or the reset pins of the latches
or counters respectively. In this way the time period for which the counters are operational
may be recorded in units of 0.01ms, but with an accuracy and repeatability of better than a
few tenths of a microsecond.

Thus when a reset address is issued by the computer, the normally-hi S input of the JK
flip-flop circuit drops to a 0. The flip-flop then has J=0, K=1 and Q=1, and waits for the next
falling edge of the clock waveform before its output can fall. This transition then initiates the

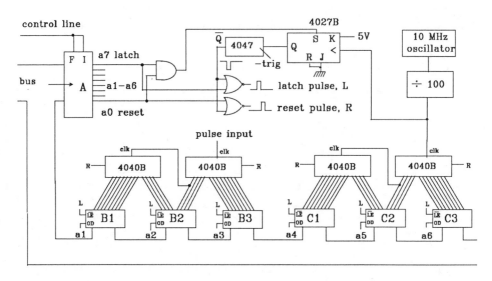

Fig 7.16 *A 24 bit pulse counter and 24 bit timer which may be zeroised, latched and read using addresses from the computer.*

output of a 1 microsecond wide pulse from the 4047 monostable to reset all of the counters to zero. Counting proceeds. When the computer needs to read the counter and timer, a latch address is issued and, after the delay required to ensure synchronisation with the clocking oscillator, the 6 bytes of data from the counter and timer are latched into the six latches. These bytes are then read by the computer one at a time in response to the appropriate address codes. If the computer is satisfied with the readings the counters may be reset to zero for a second measurement, or alternatively (as the counter and timer were left running even when the latching operation took place) a second latch address may be issued and the updated bytes transferred to the computer. This technique is particularly valuable for the counting of random pulses as it enables the data to be examined repeatedly until a required statistical error limit has been reached.

7.5 Handshaking

While the interfaces described in section 7.1 are all very straightforward, there are circumstances in which communication between any one of them and the microcomputer could be unreliable. The main problem lies in the use of a single control line to activate the interface. Using one control line operated by the computer results in the computer functioning blind - it has no way of knowing whether the interface is ready for a byte transfer. In most cases the system will work correctly under the control of a BASIC program, because BASIC is relatively slow and several milliseconds elapse between each operation connected with the interface. However, even this may not be enough time for a high resolution ADC to complete a conversion, or for a pulse output interface to discharge a specified number of pulses before receiving another instruction. The situation becomes even more serious if a compiled language or machine code is used for interface control and byte

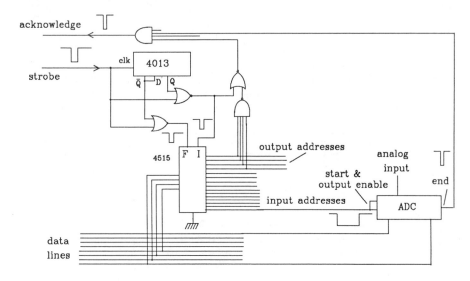

Fig 7.17   *The use of strobe and acknowledge handshaking signals
in the operation of an addressable interface.*

transfer operations, and a number of time delay loops may be required in the controlling program to prevent consecutive signals being sent to the interface too quickly.

A simple solution to this problem may be found in using two control lines, one carrying a signal from the computer to the interface (as before), and a second carrying a signal in the other direction - the handshaking technique outlined in section 5.10. The basic principle is illustrated in fig 7.17, where the strobe and acknowledge lines of a typical user port are used to carry these "handshaking" signals to an addressed (ie. multiplexed) interface. Operation of the address decoder is as described in section 7.2, and a 4515 decoder is used to generate lo levels on 1 of 16 address output lines. However the circuit also produces a pulse on the acknowledge line (ie returns a signal to the computer) whenever the required function of the interface has been completed.

The acknowledge signal may be generated from the initial strobe line pulse for those functions which are "immediate". For example, the address decoder cannot fail to latch an address from the lower four bits of the first byte received, because the circuit is hardwired to ensure that latching occurs when the strobe pulse reaches the 4515. Thus the pulse applied to the 4515's follow pin is returned to the acknowledge line and signals the computer that the address has been received. Some of the address lines in fig 7.17 (marked "output addresses") are also connected via gates to the acknowledge line. Thus if any one of these address lines is pulsed low (which occurs after their address has been latched and when a second strobe has been received) an acknowledge pulse is automatically generated. Consequently devices connected to these address lines, such as the DACs and logic outputs shown in fig 7.18, do not need to generate their own acknowledge signals to confirm that they have received data - the address decoder does that for them.

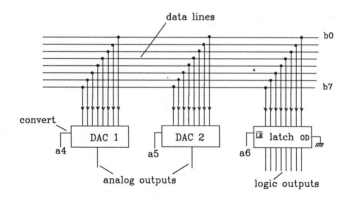

'Fig 7.18 *Simple output functions may be implemented without generating specific acknowledge signals, as these may be produced from the address decoder output address lines (see fig 7.17)*

Other address lines in fig 7.17 (those marked "input addresses") do not generate acknowledge pulses automatically. Thus the example shown in fig 7.17 illustrates a single input address line connected to the start pin of an ADC. A hi-to-lo transition on this line starts the ADC's conversion process. When the conversion is complete the ADC generates an "end of conversion" signal in the form of a pulse, and it is this pulse which is passed through the AND gate on the acknowledge line to form the acknowledgement signal. (The lo on the address line could also be used to enable the ADC's digital output so that it could be read by the computer.) The important difference is that the acknowledge pulse in this case is generated after the ADC has completed its conversion, which could be a variable time after the receipt of the start conversion signal, rather than immediately. Thus there is no danger that the computer will attempt to read the ADC's output before that output has been formed. The control line sequences associated with the output of an address code followed either by the output or input of a data byte are illustrated in fig 7.19. Note that the strobe signal required for byte input is somewhat more than a pulse, in that the strobe line remains low until after the acknowledge pulse has been received by the computer. In fact for data input the strobe and acknowledge control lines are probably better described by the names "request for data" and "data valid" respectively.

Control of the addressable handshaking interface of fig 7.17 may be achieved using BASIC routines which are simple extensions of routines 300 and 400 of section 6.3. Examples are shown in routines 500 and 600 below, although a number of assembler language routines which incorporate some error checking have been published (Laboratory Microcomputer, vol 6, 122-126, 1987).

```
500    REM output byte da% to address ad%
501    REM assumes OUT 1003, 193 has been performed
510    OUT    1001,3:            REM PB0 HIGH - ACK TO PC6
515    OUT    1000, AD% :        REM OUTPUT ADDRESS
```

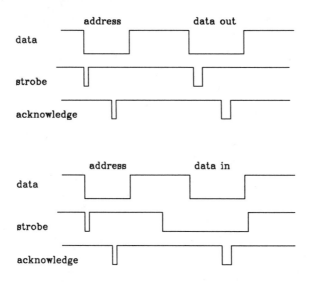

Fig 7.19 *The handshaking signals involved in addressed byte input and output using the address circuit of fig 7.17*

```
520    OUT    1001,2 : OUT 1001, 3:     REM PB0 LOW THEN HIGH
525    IF (INP(1002) AND 128)=0 THEN GOTO 525 : REM WAIT ACK PULSE
530    OUT    1000, DA% :       REM OUTPUT DATA
535    OUT    1001,2 : OUT 1001, 3:     REM PB0 LOW THEN HIGH
540    IF (INP(1002) AND 128)=0 THEN GOTO 540 : REM WAIT ACK PULSE
550    RETURN

600    rem INPUT byte TO da% FROM address ad%
601    rem  assumes OUT 1003, 193 has been performed
610    OUT    1001,3:            REM PB0 HIGH - ACK TO PC6
615    OUT    1000, AD% :        REM OUTPUT ADDRESS
620    OUT    1001,2 : OUT 1001, 3:  REM PB0 LOW THEN HIGH
625    IF (INP(1002) AND 128)=0 THEN GOTO 625 : REM WAIT ACK PULSE
630    OUT    1001,5 :           REM PB0 STIL HIGH BUT ACK TO PC4
635    OUT    1001,4 :           REM PB0 LOW - REQUEST DATA
640    IF (INP(1002) AND 32)=0 THEN GOTO 640: REM WAIT ACK
                                                    (DATA VALID)
645    DA%=INP(1000) :           REM READ DATA AND STORE IN DA%
650    OUT 1001,5 :              REM RETURN PB0 HIGH
655    RETURN
```

BASIC versions of these routines are able to handle byte transfers at up to about 5 address+ data pairs per second on a normal PC. QuickBASIC version can handle up to about 50 pairs per second, and assembler routines up to 1000 pairs per second.

7.6 Synchronous byte transfers

The byte transfer techniques we have considered so far have been asynchronous techniques, ie. addressing the interface and transferring a data byte have been operations independent of the MPU's clocking signal. This has required that the interface address and the data byte are transmitted over the port bus one after the other, and that control signal and data direction registers in the PIA are changed during each transfer operation. Consequently none of the techniques described qualifies as "fast", even though the use of machine code may allow transfer rates of a thousand bytes per second or so. However, there is another method of communication between the computer and external devices, which operates in much the same way as byte transfers between the MPU and ROM or RAM, and allows us to make use of the speed characteristic of those transfers.

As the MPU's address bus is generally available for use outside the computer (even if partially decoded), we can use this to carry addressing signals to an interface unit. Furthermore the timing signals available on the control bus (see section 5.9) provide the means of operating latches for the latching of address bus codes and for reading from or writing to the MPU's data bus. If these facilities are utilised we may transfer bytes directly between the MPU and external devices without the aid of a PIA or similar system. The availability of address bus signals and the precise control bus signals varies somewhat between one micro and another. Rather than cover all possibilities we will confine our attention to address decoding of the address lines available on the PC's expansion slots, where only a limited number of addresses (768) are available, and the IOR and IOW signal technique is used to synchronise operations.

First it is necessary to point out that if particular IO addresses are to be used to initiate data byte responses from an interface, then those addresses must not produce responses from elsewhere in the computer (eg. from other interface cards in the case of a PC). The IBM Technical Reference Manual specifies a number of IO addresses which are either unused or set assigned for user cards (eg, the "prototyping card"). We start by assuming that we have identified a block of 16 consecutive addresses which can be used to address an interface system, and that these addresses cover the range 0 - 15 on the least significant address lines, A0 - A3. Decoding A0 - A3 is, of course, quite straightforward using a 1 of 16 decoder. However, we also need to verify that this block of 8 addresses has been addressed for an IO operation by the MPU, and this requires us to check the values of 6 other address lines, A4 - A9. This is most readily achieved using a digital comparator, a logic circuit which gives a lo output if two multi-bit digital inputs are identical, and a hi output otherwise.

The 74HC688 digital comparator operates on two 8 bit inputs (called P and Q). When enabled (with a lo applied to the G input), the comparator produces a lo output when the digital values of the P and Q inputs are equal, and a hi output otherwise. The circuit of fig 7.20 illustrates how one of these devices can be used to generate a pulse output while an address (preset by the bank of switches or jumpers) is present on lines A4 - A9 of the address bus. The pulse timing is derived from the AEN signal of the expansion bus which operates the comparator's G input. The resulting lo pulse is then ORed with the IOR and IOW levels to produce a lo "read" pulse or a lo "write" pulse - depending on which of IOR or IOW went

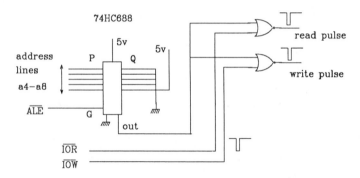

Fig 7.20 *Partial decoding of the six bits A4-A9 of the MPU address bus using a*
74HC688 digital comparator. Note that either a read pulse or a write pulse is
produced when an IO address match is present, depending on the level of the
IOR and IOW control bus line.

lo. This enables us to use the 16 available addresses for 16 input functions and 16 output functions independently.

If the part address set by the bit select jumpers in fig 7.20 corresponded to 30x, ie 110000xxxx (where the xs are the least significant 4 bits and are not used in this part of the circuit), which is equivalent to a decimal number in the range 300H - 30fH, then a read pulse would be generated by X = INP(768) etc. and a write pulse by OUT 768,X etc. (or the equivalent in and out instructions in machine code). These pulses may now be used to enable (one of) a pair of 1 of 16 decoders connected to address lines A0 - A3 as shown in fig 7.21. The enabled decoder will produce on one of its 16 output lines a pulse synchronised with the IOR or IOW signals of the expansion bus. A normally hi output decoder such as the 74HC154 (older CMOS devices would be rather slow at this point - the decoding time of the 4515B is about 800 ns) produces a negative-going pulse which can be used to operate the output disable line of one of the TRI-STATE latches connected to the MPU's data bus using the leading edge of the read pulse. This allows the data to be loaded on to the data bus in good time for latching by the computer at the end of the read pulse. In this manner X=INP(768) both selects the appropriate latch in the interface circuit and transfers the data byte into the computer (storing it into a BASIC variable, X) quite rapidly (a few microseconds if machine code is used). The trailing edge of a write pulse is used to latch data from the data bus (in line with the memory timing discussed in chapter 5), so the negative-going pulse from the decoder must be inverted as shown. Thus at the end of the pulse the latch collects the byte loaded onto the data bus by the computer's OUT 768,X instruction.

It is likely that some of the addresses decoded would be used for control functions in a complex interface system, just as they were in the dual counter/timer system of fig 7.14. Others may be used to check for busy signals by gating a read pulse with a busy signal output using a TRI-STATE latch as illustrated in fig 7.22. In this case the state of an ADC can be tested until conversion is complete, and then the data read. An example of some suitable coding is:

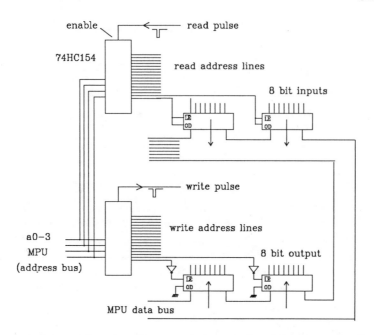

Fig 7.21 *Decoding bits 0-3 of the MPU address bus in conjunction with the read/write pulse produced in fig 7.20 allows control of 16 input and 16 output functions.*

```
100   Y=INP(768)  :            REM initiate conversion
110   REM check ADC busy signal
120   IF (INP(769)AND1)<>0 THEN 110: REM check again
130   X=INP(770)  :            REM read ADC into X
140   REM DATA BYTE NOW IN X
```

Note that the convert signal was sent by addressing 1000 using a INP(), although no useful data would be stored in Y. We could equally have used a spare write line and OUT 1000,0 to initiate conversion.

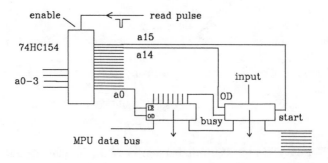

Fig 7.22 *Using a logic or sense input interface function to monitor the busy lines of other and slower devices.*

7.7 Dynamic Interfaces

The principal difficulty in using ADCs for measuring small signal changes is that the resolution with which the change is recorded may be very much poorer than the full scale resolution of the ADC. For example, a chromatographic uv absorbance detector may produce an output which can cover the analog range 0 - 2.55 V, while the chromatographic peak from a sample component may occupy the much narrower range between 0.75 and 0.8 V. Using this extreme example for simplicity, we can see that while the full scale range (0-2.55 V) could correspond to the 0 - 255 range of an 8-bit ADC, the small peak would be converted into digital values covering a range of only 5 units (with an uncertainty of 2, 1 at each extreme). One tempting solution to this problem may be to make use of a higher resolution ADC. If we apply a 12-bit ADC to our example values we find that the full scale analog range corresponds to 0 - 4095, and the peak would be converted into a digital range of about 80 units (again with an uncertainty of 2). This approach is fairly readily implemented when the number of bits is small, but becomes much more expensive as the required resolution increase beyond 16 bits. Another difficulty with increasing the resolution is that high resolution ADC tend to be relatively slow converters, although the conversion speed may be variable.

An alternative approach is to change the analog range which the ADC uses as its full scale range. For example, if the full scale range of our 8-bit ADC was 0.7 - 0.85 V, then the peak would be converted into a digital range covering 80 units, with the same uncertainty as that obtained from a 12-bit ADC operating over the wider full scale range. One technique for implementing this is illustrated in fig 7.23. An analog input signal, V_{in}, is passed to an amplifier which produces an analog output, V_a, given by

$$V_a = V_{in} - V_{os}$$

where V_{os} is an analog offset provided by a DAC.

The offset value is changed by the controlling computer until the value converted by the ADC lies within its input range. The value of V_{in} may now be calculated in the computer from

$$V_{in} = V_a + V_{os}$$

This is an example of a dynamic analog input interface, where the analog range of a simple ADC is adjusted to suit the measurement being undertaken.

At first sight this approach would appear to offer no advantage over the alternative solution suggested above. However, two advantages are available and may be deciding factors when resolutions over 12 bits are required. The first advantage lies in the fact that an 8-bit ADC and an 8-bit DAC together can cost considerably less than a 16-bit ADC alone, and the difference in cost becomes even more substantial at higher resolutions. The second advantage may lie in the speed of conversion, but requires a little thought to determine whether this advantage may be realised in a given system. It is probably the case that if V_{os}

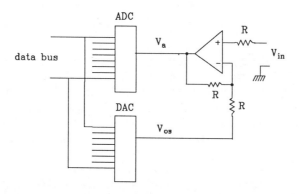

Fig 7.23. *Analog input interface which converts the voltage difference between the input voltage and an offset generated by a DAC.*

required changing for every reading of V_{in}, then the speed advantage would lie with a higher resolution ADC. However, when small signal level changes are being recorded it may be possible to set up the dynamic analog input interface so that the signal may be converted without V_{os} having to be altered. In such cases the smaller resolution ADC used in the dynamic interface will be considerably faster than the higher resolution ADC required in a simple interface.

It should be noted that it is difficult to achieve the simulation of 16-bit resolution by using a dynamic interface for which the total number of information bits (ie ADC + DAC) available is 16, because some overlap of ranges is virtually inevitable and probably desirable if difficulties in software control are to be avoided. In practice it seems sensible to design such dynamic interfaces to operate with at least 1-bit redundancy, so that an 8-bit ADC may be combined with an 8-bit DAC to provide data as valuable as a 15-bit ADC.

Dynamic analog input interfaces may also use adjustable gain amplifiers to change the full scale analog range, as illustrated in fig 7.24. Indeed there are a number of instrumentation amplifiers available with gains which may be set by a pattern of logic levels (such as the LH0084 from National Semiconductor). In fig 7.24 a more mundane method of adjusting the gain is used, involving the selection of one of four feedback resistors for a non-inverting op-amp using an analog switch. Assuming that an ADC with an input range of 0 - 2.56 V is used, V_{in} may be calculated from

$$V_{in} = V_a / G_i$$

where V_a is the voltage measured at the ADC input and G_i is the gain selected.

Again the selection of the conversion gain by software can allow a changing signal to be converted under favourable conditions, and again a speed advantage may be achieved if it is possible to operate the measuring system in such a way that gain changes during the recording are not required. The circuit elements introduced in figs 7.23 and 7.24 may be combined to provide a dynamic interface in which both the gain and offset may be set by the computer. In a sense, setting the conversion gain or offset for a dynamic analog interface is

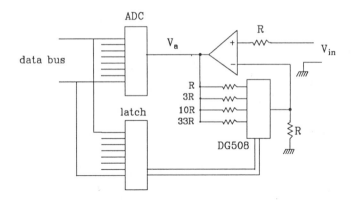

Fig 7.24. *Analog input interface which amplifies the input voltage using a computer-selected gain.*

analogous to selecting the range and zero position of a chart recorder; the setting may be arranged to provide a recorded signal of optimum value, but having to reset parameters for signal variations which are too small or too large can be an inconvenience.

In practice we have applied selectable gain amplifiers to signal conditioning circuits and confined the dynamic action of our analog input interface to the variable offset variety of fig 7.23, although using a fixed gain of greater than unity. A useful circuit for chromatography and flow analysis applications is shown in fig 7.25, based on a ZN428 DAC and a ZN448 ADC. Both converters share the same reference voltage derived from the ZN428 and nominally 2.55V. The 4528 dual monostable is present to provide a "data ready" reply for the computer when it addresses the ADC. One of the monostables is triggered by the "conversion complete" edge from pin 1 of the ADC.

The difference between the analog input signal and the DAC output is multiplied by (nominally) 47 by the OP07 amplifier, which is operated as a high impedance input (33 M) non-inverting amplifier as far as the analog input signal is concerned. This amplified difference signal is passed to the ADC for conversion. Unfortunately the ZN448 does not like receiving a negative input signal or one of more than 3 V; higher input voltages cause the ADC to produce digital values smaller than 255. As this cannot be detected by software, we added the diodes between the analog input pin (pin 6) and ground to ensure that the input voltage could not exceed limits of 3 V (set by a 5 V Zener) and -0.6 V (set by a silicon signal diode). At these voltages the ADC correctly returns 255 and 0 respectively, allowing the software to detect that the signal level is at the end of the range.

Examples of 6502 and 8086 assembler language routines used to operate this dynamic interface have been published (D J Malcolme-Lawes, Laboratory Microcomputer, 6, 3, 122), although for low speed applications in which the signal changes very slowly (ie at less than 1 least significant bit per second) it is possible to operate the interface using compiled BASIC. The BASIC routine listed below indicates the principles of operation, including the handshaking through the user port adaptor.

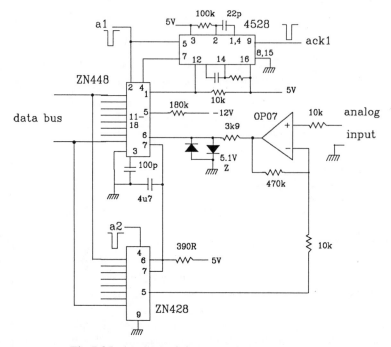

Fig 7.25. *A practical dynamic analog input interface.*

The calling program merely calls the interface handling subroutines and then evaluates the analog value from an equation of the type:

$$V_{in} = V_{os} - V_a/G$$

where G is the amplifier's gain, nominally 47, which requires accurate evaluation. There is only one unknown, G, and this may be readily determined by applying an accurately measured signal level to the input and noting the ADC and DAC values returned. We have found that it is only necessary to use one signal level from each end of the range, typically 0.1 and 2.4 V, to ensure adequate calibration, as the OP07 op-amp has a very low differential input offset voltage.

```
1          DEF SEG
2          GOSUB 60000
10         GAIN = 47.135
           ......
300        REM get current adc reading adjust dac auto
310        CALL DANALOG (ADC%,DAC%,ER%)
320        IF (ER%=255) THEN VADC=-1:RETURN
330        IF (ER%=0) THEN 350
340        PRINT"OUT OF RANGE": RETURN
350        VADC=DAC%+ADC%/GAIN
```

```
360        PRINT VADC;" volts"
399        RETURN
           ......
60000      REM
60010      OPEN "routs.exe" FOR INPUT AS 1
60020      FOR I=1 TO 4: IN$=INPUT$ (128,1): NEXT I
60030      DOUT=60000: DIN=60002: DANALOG=60004
60040      FOR I=1 TO 360: IN$=INPUT$(1,1): POKE DOUT+I-1,ASC(IN$): NEXT I
60080      PRINT"code loaded"
60085      CLOSE 1
60086      STOP
60090      RETURN
```

Not surprisingly it is possible to obtain commercial dynamic analog input interfaces with considerably better performance. If a very wide dynamic range and high accuracy are important, then one of the autoranging multidigit digital voltmeters fitted with a standard interface system for connection to a computer offers a versatile and reliable solution. While the dynamic interfacing described above was illustrated using an 8 bit ADC, in practice a higher resolution ADC would be more appropriate. High quality high resolution ADCs are moderately expensive and in systems where several analog inputs are to be monitored it is generally desirable to switch the analog input to the ADC from several sources, rather than have an ensemble of separate ADCs monitoring all the inputs. Switching analog inputs can be carried out using analog multiplexor ICs, but there are a number of ADC circuits on the market with built-in multiplexors. Several of the multichannel ADCs are intended for direct connection with microprocessors and are equipped with MPU compatible address, data and control lines (some are able to provide signals to generate MPU interrupts at the end of conversion). We will consider one such device, although discussing its connection to a microcomputer using the types of addressing described earlier in this chapter.

The part circuit of fig 7.26 shows the HS9410 (Hybrid Systems Corporation) 8 channel, 12 bit ADC connected as an analog input interface using the synchronised addressing method. Any of the other methods would be equally suitable for the limited use we make of this device's facilities, as the (12 bit) conversion time is about 30 microseconds. The device requires 5 inputs: 3 address lines to carry a code for selecting 1 of the 8 analog inputs; a "read/convert" line (R/C), which enables the digital output when hi, disconnects the digital output when lo (ie. the output goes to the high impedance state), and initiates the ADC conversion on a hi-to-lo transition; and a "device address" line (A) which selects 12 bit conversion if lo and 8 bit conversion if hi when R/C receives a hi-to-lo transition, and selects which byte is output on the data lines when R/C is hi. (A lo gives the most significant byte. A hi gives the least significant byte - including 4 trailing zeros.)

In this implementation the circuit is addressed using the synchronous technique illustrated in figs 7.20 and 7.21. The 3 bit select code and the 1 bit conversion type code are provided as data via latch A using OUT a0,X. Conversion is initiated by addressing a1 (ie. Y=INP(a1)), and data may be read from latch B (address a2) either as one byte (for 8 bit

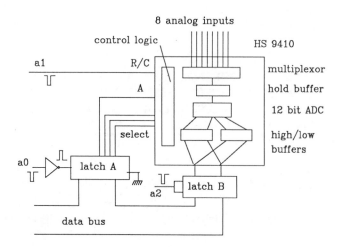

Fig 7.26 *A multiplexed input analog input interface allowing the reading of up to*
          8 analog signals with a 12 bit ADC.

conversions) or as two bytes selected by modifying the bit value of A. The following BASIC
commands illustrate the procedure:

```
10    A0=768: A1=769: A2=770: REM ADDRESSES
20    ABYTE=N+0: REM for 12 bit conversion, channel N - N+8 for 8 bit conversion
30    OUT A0,ABYTE: REM SELECT CHANNEL
40    T=INP(A1) : REM INITIATE CONVERSION
50    REM MACHINE CODE ROUTINES REQUIRE A TIME DELAY
                        TO ALLOW FOR CONVERSION
60    ABYTE=ABYTE AND 7: OUT A0,ABYTE: REM SELECT HIGH BYTE
70    H=INP(A2): REM READ HIGH BYTE
80    ABYTE=ABYTE OR 8: OUT A0,ABYTE: REM SELECT LOW BYTE
90    L=INP(A2): REM READ LOW BYTE
100   VALUE= (256*H + L)/16 : REM EQUIVALENT 12 BIT VALUE
```

If the equivalent assembler language instructions are used to implement these operations then
a delay is required between 12 bit initiation and the high byte read (about 30
microseconds), and between the 8 bit initiation and the single byte read (about 21
microseconds). The delay may be programmed with a suitable NOP loop. The
manufacturer's data sheets shows how conversions may be "overlapped" to achieve a very
high data acquisition rate.

CHAPTER 8

PARALLEL STANDARD INTERFACE SYSTEMS

Interface standards exist so that units from one manufacturer can be connected to devices from another manufacturer and still allow digital signals to be passed from one to the other. In the computer context a standard interface system can be viewed as consisting of the elements illustrated in fig 8.1. The "system" actually consists of two interface units - the computer interface and the device interface. To allow for digital signal transfer between the two, the "standards" are primarily concerned with the characteristics of the signals involved - their number, signal levels, impedances, timings etc. - rather than with the actual electronic circuits in either interface, although some characteristics of the latter may be important.

8.1 Introduction

A growing number of standard interface systems is available, but it is important to understand that some of the most widely used standards were not actually devised for the convenience of the microcomputer users (or manufacturers). Some were developed for use in technical instrumentation, to allow one piece of electronics to control another, etc. Others are primarily communications standards, developed for the interconnection of communications equipment (eg. terminals to modems). Those that have been adopted in the microcomputer world are not necessarily ideally suited to the task, but they have dramatically increased the number of devices that can be "plugged in" to suitable microcomputers. They have also reinforced the value of the standards to the extent that most microcomputer peripherals and many laboratory instruments now incorporate a device interface conforming to one of the major standards. In this chapter we will examine two of the most popular standard interface

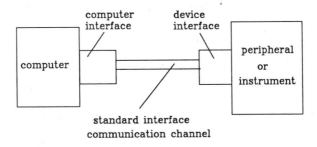

Fig 8.1 *The elements of a standard interface system.*

systems, one (the Centronics parallel output interface) designed primarily for the connection of printers to computers, and one (the IEEE 488 standard) intended for use with laboratory instrumentation. In the next chapter we examine a serial interface standard (RS232) primarily designed as a standard for communications equipment - but still one of the most widely used micro input/output standards. Many microcomputers are fitted with one or more of these standard computer interfaces, or may have such interfaces added to their hardware. The advantages of standard interface systems are that

a) if one buys a new computer it is not necessary to buy a whole new range of interfaces for it - only one standard computer interface is required,
b) a large range of instrumentation with standard device interfaces built in is available from dozens of different manufacturers, and
c) in some cases the program control of the standard interface system can be exercised by normal file handling commands from high level languages such as BASIC (ie. no INP or OUT routines are required).

Against these must be weighed the disadvantage of the high cost of some standard device interface on simple devices (such as analog input or pulse counter interfaces) compared with the simple parallel port system described in chapter 7. Even for in-house interfaces, the additional complexity of the circuitry required to implement standard interface specifications can be considerable.

8.2 The Centronics standard

The Centronics standard interface was designed to enable printers to be connected to computers (Centronics Corporation is one of the major manufacturers of printers). Fortunately it is a standard which has been adopted and largely adhered to by most other printer manufacturers throughout the world. It offers the advantage of a relatively fast byte output interface available on the majority of microcomputers, and in most cases the software routines for handling the byte handshaking is provided by high level languages, the micro's operating system or its ROM routines. Although initially created as an output-only interface, bidirectional Centronics interfaces are now quite common, often on peripherals which are associated with relatively large numbers of byte transfers (such as optical scanners). Essentially the standard is based on three groups of connections between the computer and the peripheral: an 8 bit parallel data connection, a two-wire handshaking connection, and a number of connections used for control signals and error detection. These connections are illustrated in fig 8.2.

Strictly speaking the standard specifies the connector on a printer (a 36-way Amphenol connector) and the signals required by and provided by the printer. For this reason there is a considerable amount of variation in the connector provided at the computer end, although most manufacturers now use either a 36-way Amphenol connector or a 25-way D connector (known as a DB25). All the connectors on equipment are sockets - the connecting leads are fitted with plugs. The standard connections are detailed in fig 8.3, where a PC's DB25 connector is shown at the computer end to illustrate the translation of the pin numbering when an insulation displacement (IDC) ribbon cable is use. The pin configurations adopted

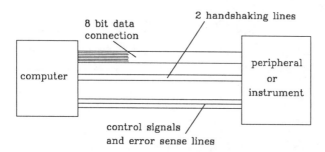

Fig 8.2 The principal connections of the Centronics standard interface.

on parallel printer ports by IBM (for the PC) and Commodore (for the Amiga) are shown in table 8.1. While a correctly wired cable is the ideal, in practice a one-to-one DB25 to Amphenol 36-way connector (such as may be readily produced using IDC systems) will generally function for a PC system if the line to DB25 pin 15 (the fault line) is broken, although error conditions may be reported as "printer busy" by some software. Some companies produce IDC connector specifically to allow PCs to be connected to Centronics devices - although naturally these cost more than standard connectors.

Nine of the pins on the 36-way connectors (14-18 and 33-36) are specified as not normally used, so, naturally, different manufacturers have used these pins for different purposes. Unfortunately one of the main uses which some manufacturers have found for at least one of the spare pins (often pin 18) is to carry a 5 V power level. At one time this was a useful facility for providing power from a printer to an interface converter, but now it has

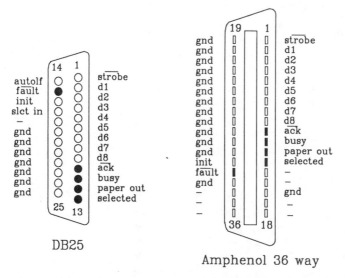

Fig 8.3 *The pin connections commonly used for 25-way (left) and 36-way (right) connections to Centronics standard interfaces.*

Table 8.1   Parallel printer port connections

Centronics function	Centronics 36-way	IBM PC 25-way	Amiga 25-way	other use
-strobe	1	1	1	
data 1	2	2	2	
data 2	3	3	3	
data 3	4	4	4	
data 4	5	5	5	
data 5	6	6	6	
data 6	7	7	7	
data 7	8	8	8	
data 8	9	9	9	
-acknowledge	10	10	10	
busy	11	11	11	
paper out	12	12	12	
select	13	13	13	
		14		-auto lf
-input prime	31	16	25	
-fault	32	15		
	36	17		-online
ground	19-30	18-25	14-22	ground
			23	(+5 V)

become something of a hazard as it is all too easy to connect to another device on which pin 18 is grounded. Another example of this difficulty is apparent in a comparison between the 25-way connectors provided on the IBM PC and the Amiga. In one case pin 23 is grounded, in the other it is a 5 V power line. Pin 25 of a DB25 plug wired using insulation displacement ribbon cable would coincide with pin 19 of a 36-way Amphenol. In other words, when connecting an Amiga to a Centronics printer make sure you have the correct cable for the job, or the computer's 5 V line may get grounded!

    Data is transferred from the computer to the printer in byte serial - bit parallel form (ie 8 bits at a time), with handshaking provided by negative- going strobe and acknowledge pulses as illustrated in fig 8.4. The computer places the data byte on the data lines, waits a period of time for the data to become stable, then transmits the strobe pulse - at which point the printer should latch the data value. As the latching is generally accomplished using a circuit having a latch enable function, the data is normally latched on the rising edge of the strobe pulse. The computer is then free to remove the data byte from the data lines - although some implementations do not remove the byte until after the acknowledge signal has been received. The printer then processes the data byte (eg. prints it or stores it in a buffer) and then issues an acknowledge pulse to the computer to indicate that has finished with that data transfer. The timings specified for the stabilisation time and pulse widths are indicated in fig 8.5, although many printer manufacturers (particularly those of small battery-powered printers)

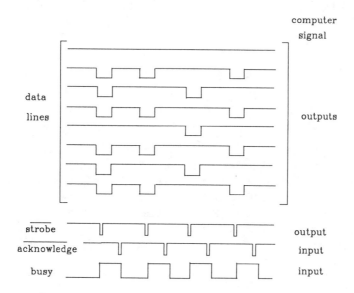

Fig 8.4 *The transfer of byte-serial data via a Centronics standard interface channel.*

appear to adopt quite different values, some requiring strobe widths of more than 10 ms and generating acknowledge pulses wider than 80 ms. The standard also specifies that a busy signal should be provided by the printer when it cannot accept new data, including during the interval between the strobe pulse and the end of the acknowledge pulse. Because the signals are negative acknowledge and (positive) busy, both rising (on acknowledge) and falling (on busy) edges are available at the end of each transfer. Different computer systems have made use of different edges to initiate their "transfer complete" actions, and this has resulted in some confusion about which of the two connections (busy or acknowledge) should be passed to the computer's handshaking connection.

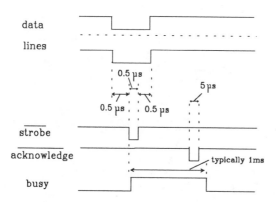

Fig 8.5 *Typical handshaking timings associated with data transfer via Centronics standard interface channel.*

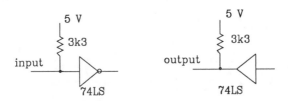

Fig 8.6 *Typical elecronic configuration of device interface circuits used for Centronics standard interface channel.*

The Centronics printer interface is implemented using circuitry which should be driven by open-collector TTL outputs (ie the printer inputs are pulled-up by resistors as illustrated in fig 8.6), although most Centronics-compatible devices will work when driven by tristate circuits. It is because an open collector output buffer is usually provided in the computer's interface that it is not always possible to reprogram the interface circuitry to operate as a parallel input interface. Thus most PC clones cannot have their printer ports reprogrammed for input. On the other hand the parallel port which normally provides the Centronics-compatible output interface on the Amiga may be reprogrammed to handshake data bytes on input.

The user port adaptor for the PC introduced in chapter 6 may be used to drive a Centronics-compatible peripheral as illustrated in fig 8.7 - and in fact the connections of the 8255 may be reworked to provide this facility more efficiently (although at the cost of losing

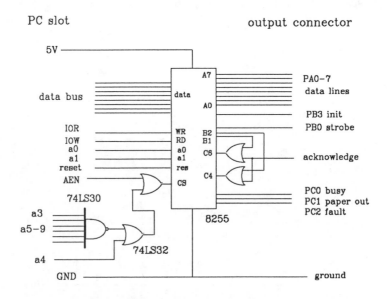

Fig 8.7 *The PC user port adaptor of chapter 6 with spare lines (PB3, PC1 & PC2) handling Centronics control signals.*

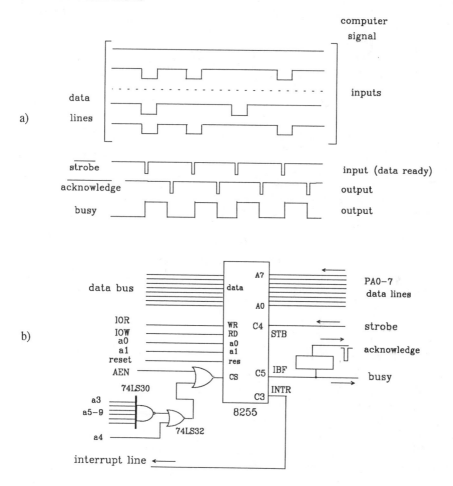

Fig 8.8 *Parallel interface data transfers to a computer. a) shows the handshaking of data bytes, and b) illustrates one implementation of the computer interface using an 8255.*

the user port compatibility) as described in the Intel data sheet for the 8255. Conversely the output functions of the interfaces described in chapter 7 may be operated from a standard Centronic-compatible interface, although to make use of operating system software (eg. device LPTn: in MS-DOS) to drive such interfaces it may be necessary to tie several of the input control lines (such as busy, paper out, etc) to fixed logic levels to prevent the software reporting error conditions.

The relative simplicity and efficiency of the parallel data handshaking transfer technique has led to the increasing popularity of a similar approach for data input to a computer. It must be realised that there is a more fundamental difference between input and output than merely the direction of the data transfer. The more important question is: who decides when it is time to transfer the data? In the case of the user port interfaces introduced in chapter 7 all data transfers, input and output, were initiated by the computer - by a low on the strobe line. When a computer initiates an input byte transfer in this way it (the computer) must wait for the peripheral to signal that the byte is available on the data lines before the byte can be read. However, in many applications, particularly those in which a large number of bytes may be transferred, it is more efficient to have the peripheral device signal when data is ready and the computer respond by reading the byte and acknowledging completion.

An example of the handshaking involved for this approach is illustrated in fig 8.8a, while a circuit which implements the interface using an 8255 is shown in fig 8.8b. The 8255 is operated in mode 1 (control byte = 176). A lo on the strobe input latches the data into the input buffer of port A and generates an input-buffer-full signal which acts as a busy signal (the rising edge of this signal may also be used to generate an acknowledge pulse although the components for this are not shown). The 8255 then generates an interrupt request to the host computer, and the interrupt service routine may read port A - a process which automatically clears the input-buffer-full signal (ie the busy line). Note that most parallel interface cards for PCs cannot be configured to input data, so this technique generally requires the installation of a purpose-designed card. Devices which transfer data in this way are generally accompanied by such a card and a suitable software device driver.

8.3 The IEEE 488 standard

The best known standard interface system for use with laboratory instrumentation is the IEEE 488 standard, developed originally by Hewlett Packard and often known as the Hewlett Packard Interface Bus (HPIB) system by Hewlett Packard or the General Purpose Interface Bus (GPIB) by everybody else. (In its present form the standard is IEEE Std 488-1978. The full specifications are available from: IEEE SERVICE CENTER, 445 Hoes Lane, Piscataway, New Jersey 08854, USA, and helpful documentation is available from Hewlett Packard. This standard is identical with ANSI MC1.1.). The standard covers the form of data transmission (8 bit parallel, byte serial), the maximum byte transmission rate (1 Megabyte per second - although no microcomputers can actually manage that), the signal levels (TTL compatible), the length and number of wires used for data lines and control lines, even the stackable connectors used on the 24-way cable between instruments. (A closely related standard is the IEC 625-1, which calls for a different style of plug and 25-way cables between units. Adaptors are available for interconnecting the two systems, from N.V. Philips and A.G. Siemens among others.) As a small aside it is interesting to note one of the problems encountered by international standards. The stackable connector of the GPIB uses pillar bolts to secure the mechanical connection. Some of these standard connectors have black bolts, and these have metric threads. Other connectors have silver-coloured bolts, and these have imperial threads. If these two different threads are inadvertently screwed together they tend to become jammed and to damage each other's threads.

A full description of the GPIB system is beyond the scope of this text, but a survey of those aspects most apparent to the user is included because some computers (eg. Hewlett Packard models) have I/O ports which are compatible with the IEEE 488 standard and software to handle data transfers, while others (eg. the Apples and PCs) can have IEEE 488 interface systems added at relatively low cost. A number of manufacturers now produce

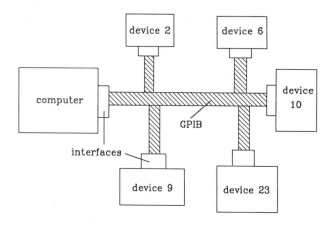

Fig 8.9 *The connection of several devices to a GPIB. Devices are addressed using a device address and only the addressed devices talk or listen.*

GPIB-compatible interfaces of the types described earlier, and a range of instrumentation fitted with GPIB interfaces, from digital multimeters and signal generators to storage oscilloscopes and graph plotters, is available from a large array of manufacturers. The most important feature of the GPIB system is that up to 15 device interfaces may be connected to a common bus over which data and commands may be passed using a byte serial bit parallel transmission technique. An example is illustrated for a typical system in fig 8.9, although where a large number of devices is involved it is better if the devices are connected by "daisy chaining" in order to minimise the capacitance of the bus wiring. Most GPIB inter-unit cables are 1 or 2 metres in length, and the standard specifies a maximum length of 4 m (although 8 m cables are available!) and a maximum total length of cabling of 20 m.

Communication over the bus is controlled by one of the devices connected to the bus and known as the Controller. Although other arrangements are possible, when one of the popular microcomputers is used to communicate with a number of devices via the GPIB, the microcomputer is used as the controller of the bus, and generally is the only controller which may be connected to the bus - thus it is not usually a simple matter to connect a number of micros together using the GPIB. (These limitations arise because many microcomputer implementations of the GPIB allow only for a subset of the standard communication signals.) When the computer needs to communicate with a specific device attached to the bus a device address is used, each interfaced device being referred to by a device address which can be selected by means of switches on the interface (sometimes not easily located or altered in low cost units). Only the addressed device interface responds to data or requests for data from the controlling computer.

Of fundamental importance in understanding communication over the GPIB is that all signals transmitted on the bus are based on negative-true logic, including those signals which represent data bytes. Thus a lo level is to interpreted as a 1, and a hi level as a 0. This relatively unusual arrangement will soon become apparent as we discuss the principal features of the bus. All lines idle at the hi level and all devices connected to the bus must allow other devices to assert a lo level on any line. Thus while bus electrical signals are all TTL-type, connections to the bus may only be made via open collector or TRI-STATE outputs. In practice a number of bus transceiver circuit ICs are available (such as the 3447P, which couples directly to the 68488P interface adaptor IC), and most manufacturers now incorporate one or other of these, so the uniformity of the electrical characteristics of GPIB devices has improved considerably in recent years.

Interfaces which can only place data bytes on to the bus (eg. a simple ADC) are called Talkers in GPIB parlance, whereas those which can only accept data bytes from the bus (eg. a simple DAC) are called Listeners. More sophisticated devices which have the ability to act as both Talkers and Listeners are sometimes called Intelligent (although their intelligence may be somewhat limited). A typical example is the Hewlett Packard model 3455A multimeter, which may have its operating function (eg. voltage, current, resistance measurement, etc.) and range (eg. 100 mV, 1 V, 10 V etc.) set with bytes from the computer, and can output readings to the computer in the form of bytes representing ASCII coded characters (eg. 55.20, 1.2345 V, etc.). Of course a controlling computer may act as a Talker or Listener in addition to being the Controller of the bus.

8.3.1 The bus conductors

The interdevice connection of the GPIB is illustrated in fig 8.10. The bus actually consists of three sub-busses: a data bus of eight conductors (dio1 - dio8) plus one ground, a three-conductor control bus which deals with the handshaking of bytes along the data bus (plus three associated ground conductors), and a five-conductor management bus used for passing control signals between devices on the bus, and for differentiating between the function of bytes being passed on the data bus (only three of the management bus lines are provided with their own ground conductors). All bus signal levels are TTL compatible. The control bus lines are known as "Not Ready For Data" (NRFD), "DAta Valid" (DAV) and "Not Data ACcepted" (NDAC). These names, some of which tend to trip rather than roll off the tongue, help to remind us that the GPIB system is one of the important instances of negative logic or lo=true logic. Thus when the NRFD conductor is held lo by anything on the bus, the bus is not ready for the transmission of data. Only when the NRFD line is hi can data transmission start. The control bus conductors must be connected (and must be operational) before any byte transfer can take place over the GPIB. In section 8.3.2 we shall discuss the role of the signals handled by the control bus.

The five management bus lines are also known by abbreviations:

ATN (attention!) The Controller sets this line lo to indicate to all devices on the GPIB that the data bus is carrying an address or a command byte (as distinct from a data byte).

IFC (interface clear) The Controller sets this line lo to force devices attached to the bus to leave the bus in a standardised state (usually with all attached devices idle and waiting for a signal from the Controller).

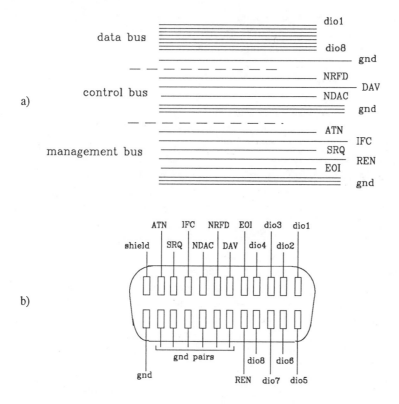

data bus

control bus

management bus

a)

b)

Fig 8.10   *a) The sub-busses of the GPIB and the mnemonics used to identify each conductor.*
b) The IEEE standard connector ribbon assignments (device socket view).

SRQ (service request) Any GPIB device fitted with the service request function can set this line lo to indicate to the Controller that it needs servicing - eg. it wants the Controller to communicate with it.

REN (remote enable) Used by the Controller to switch instruments which have the ability to be controllers to "remote" (ie. GPIB) control.

EOI (end or identify) The "end" function is used (optionally) by Talker devices to indicate that a data byte being transferred is the last data byte of a sequence, the EOI line being held lo during this byte transfer. The "identify" function is somewhat rarer, and is used by the Controller during a parallel poll (a sort of roll call in which up to eight devices can signal their presence by holding individual data lines lo) of devices present on the bus.

It is possible to use a GPIB connector on a micro to communicate with a device interface without any connection to the management bus. However, if several devices are connected to the bus it is usually desirable to operate them by addressing the individual units, which in turn requires that the ATN line is used. ATN and the three control bus lines (NRFD, NDAC and DAV) form a set which can be regarded as the minimum configuration for controlling the transfer of data over the data bus. Some micros equipped with GPIB interfaces do not monitor or use all of the management bus lines, and several commercial GPIB device interfaces intended for use with microcomputer peripherals implement only ATN and EOI.

8.3.2 The byte handshaking sequence

The byte handshaking sequence illustrated in fig 8.11 occurs for every byte transferred, whether it be a data byte or an address or command byte. The sequence begins with the source device (which may be a Talker or Controller) allowing the DAV line to go hi (remember the GPIB uses negative logic, so hi is false and in this case indicates that data on the data bus is not valid) and all acceptor devices on the bus holding the NDAC line lo - indicating that they have not accepted the current content of the data lines. The source device then places the byte on the data bus and waits for all the devices on the bus to indicate that they are ready for data by allowing the NRFD line to go hi. When the NRFD line is hi the source forces the DAV line lo, to indicate that valid data is on the data bus and ready for collection, and all acceptor devices on the bus respond by holding NRFD lo - indicating that they are no longer ready for new data. There is then a time interval during which the acceptor devices have the opportunity to read the data lines. All devices should respond by releasing the NDAC line, so that it goes hi when the last device (ie. the slowest device) releases it. When this happens the source releases DAV so that it returns hi, indicating that the data lines no longer hold valid data, and the acceptor devices respond by holding the NDAC line lo. At this point one byte has been transferred from the source to all acceptor devices on the bus. In fig 8.10 the broken lines illustrate different response times which occur when several acceptors are connected to the bus.

A couple of points need to be noted in connection with the handshaking sequence. Firstly, all devices connected to the GPIB are required to permit the handshaking of the bus whether they have been addressed or not (see below). This means that the devices must either participate in the handshaking, or must allow the lines to have their levels determined by other devices. The IEEE 488 standard does not specify the precise circuits used to send or

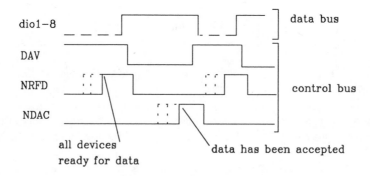

Fig. 8.11 *Level changes on the control bus during the handshaking sequence of the GPIB. The vertical broken lines indicate the different response times of devices on the bus.*

receive signals over the bus conductors, and in practice some GPIB devices use TRI-STATE circuits while others use open collector circuits whose outputs are connected via resistors to ground and a +5 V supply. The two types are distinguished by the GPIB implementation codes E1, for open collectors, and E2 for TRI-STATE buffers - which are capable of faster operation, typically 0.5 Megabytes per second. Even so, there has been a certain amount of variation in the circuits used to implement GPIB connections. A consequence of this is that some GPIB devices may not allow bus lines to return to 5 V unless the power to that device is switched on, so that having one or more unpowered devices connected to a bus can prevent the bus being used. This can result in a certain amount of confusion as other devices can be left connected to the bus while unpowered and allow data transfer between other devices to continue undisturbed. With properly designed circuits a bus should operate if at least two thirds of the devices connected are powered.

Secondly, devices differ in the amount of time they allow other devices to respond to signals on the ATN or the control bus lines. On the early PET computers, for example (the first low-cost microcomputers to implement a form of GPIB), 14 microseconds were allowed for devices to respond (by setting NRFD and NDAC lines hi and lo respectively) to a lo on the ATN line, while the IEEE 488 standard specifies only 200 ns. Similarly the PET allowed 64 ms for a device to respond to DAV lo with a NDAC hi, whereas other Controllers allow different times - some shorter, others being prepared to wait indefinitely. The timing requirements of many possible bus activites are too complex for detailed discussion in the present context, but the user should be aware that some microcomputers are actually slow devices by GPIB standards so that their use with some fast GPIB devices can give rise to difficulty. The best advice is test before you buy, as manufacturers tend not to be interested if one of their products will not communicate with a device from another manufacturer. Note that there is no restriction on the time allowed for NRFD to go hi (except in response to ATN lo), and the NRFD line is therefore used to hold up data transfers when an acceptor is busy.

8.3.3 Controller signals

The bus Controller, which is the microcomputer in the present context, always initiates the transfer of data bytes over the bus. This is true whether the computer is to act as the Talker (ie. sending the data bytes) or the Listener (ie. receiving them), or even when the computer takes no further interest in the transfer - such as when the bytes are to be sent from an instrument directly to a printer or a disk drive. The Controller initiates data transfer by taking the ATN line lo and, when all other devices on the bus have responded by letting NRFD go hi and NDAC go lo, placing a byte on the data lines. The Controller acts as a source during this operation, and when ATN is held low all other devices on the bus act as acceptors. The byte (or rather bits 0-6, as bit 7 is not used while ATN is low) placed on the bus using the convention that hi=0, lo=1, can be one of six types of "command" to the interfaces/devices connected to the bus:

A bus command. Bits 5 and 6 of the byte are 0s, and bits 0-4 carry a code (0 - 31) which the devices on the bus understand, eg. a GET (Group Execute Trigger) command - which should not be confused with a BASIC GET, see section 8.2.5.

MLA (My Listen Address, also called LAG, Listen Address Group) Bit 6=0, bit 5=1. Bits 0-4 carry an address (0- 30, not 31) which activates the device(s) with the appropriate address so that it becomes an active Listener, ready to accept data from the bus. Note that all devices handshake data bytes, but only the addressed Listener(s) actually read them.

UNL (Unlisten) The byte X0111111 causes all active Listeners to be deactivated.

MTA (My Talk Address, also called TAG, Talk Address Group) Bit 6=1, bit 5=0. Bits 0-4 carry an address (0-30, not 31) which activates the device(s) with the appropriate address to become an active Talker, ready to transmit data over the bus. The activation of a Talker automatically causes the deactivation of any previously active Talker.

UNT (Untalk) The byte X1011111 causes the active Talker to be deactivated.

MSA (My Secondary Address, also called SCG, Secondary Address Group) Bits 5 and 6 are 1s and bits 0-4 carry a secondary address (0- 31), a code which may be used for the transfer of a command to the addressed device (such as the mode in which it should operate, or, in the

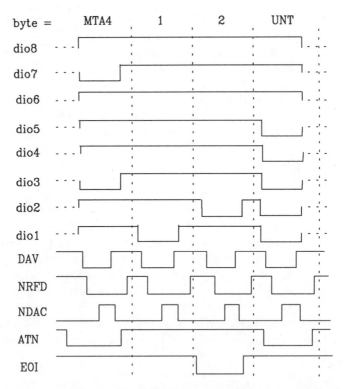

Fig 8.12 *A data transfer initiated by the Controller and carried out by the addressed talker. At the end of the transfer the Controller issues an Untalk instruction which releases the talker.*

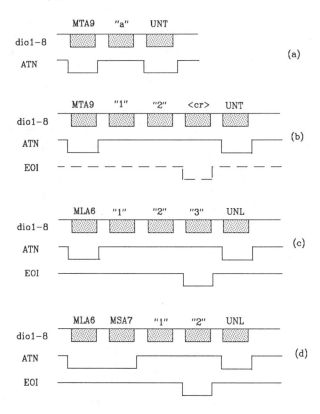

Fig 8.13 *Examples of data transfers over the GPIB. In each case the transfer is initiated by the controller issuing a MxA command, and completed by the controller issuing a UNx command.*

case of a printer, the typeface or colour with which it should print, etc.). Relatively few of the low cost GPIB devices make use of secondary address facility, although Commodore peripherals for the old PET computers used them extensively.

When the Controller wishes to initiate a data transfer sequence it outputs a MLA or MTA code (depending on whether the device addressed is required to act as a Talker or a Listener) on the data bus while holding the ATN line of the Management bus lo. For example, initiating data output from a Talker device with address 4 requires the Controller to output the byte 01000100 with ATN lo as shown on the left hand side of fig 8.12. The Controller may then either return ATN hi so that data transfer can begin, or may retain ATN lo while it outputs further bytes containing perhaps an MSA code or an MLA code (if the data is to be received by some other Listener device), before returning ATN hi. When ATN returns hi the addressed Talker device begins to handshake data bytes onto the data bus, again using the control bus to control the handshaking as illustrated in fig. 8.12, where the Talker transmits byte values of 1 and 2. The broken lines indicate that the level on the data lines is irrelevant, the actual states will depend on the type of bus drivers used.

During the handshaking of the last data byte the Talker may set EOI lo, although not all GPIB devices do this - so if a micro expects EOI to be lo to terminate the transfer of a multi-byte string of data then this could lead to trouble (I am not aware of any micros or adaptor cards which actually require this). At the end of the data transfer sequence the bus remains idle until the Controller initiates some new action. In most cases the controlling microcomputer has been either the Listener or the Talker during the data transfer, and it is usual for the Controller to deactivate the other device(s) as soon as the transfer is over. The Controller does this by issuing an unlisten (UNL) or untalk (UNT) command (depending on the nature of the completed data transfer) by holding ATN lo and handshaking 00111111 or 01011111 respectively over the bus. However, not all microcomputer (or minicomputer) Controllers do this at the end of every data transfer (the IEEE standard does not lay down rules of behaviour for data transfers) and this is one of the areas of difficulty when connecting devices from different manufacturers. Examples of some input and output transfers are illustrated in fig 8.13. Note the presence of the secondary address group command in (d).

8.3.4 Bus commands

Bytes transmitted over the bus while ATN is lo and having bits 5-7 each zero are interpreted as bus commands. These may permit the controller to find out what device addresses are present on the bus, or permit the controller to instruct devices to change their state. Commands fall into two groups, the Addressed Command Group (ACG) or the Universal Command Group (UCG). The ACG commands are issued to one or more devices which have previously been addressed as Listeners and are:

Go To Local (GTL, code 1) causes a group of devices to return to local (ie front panel) control.
Selective Device Clear (SDC, code 4) causes a group of devices to clear (ie revert to some
      predetermined state - usually the power-up state).
Parallel Poll Configure (PPC, code 5) causes a device to adjust itself so that it responds to a
      parallel poll using a specific data line when it requires service.
Group Execute Trigger (GET, code 8) causes a group of devices to take some simultaneous
      action - such as start taking measurements.
Take Control (TCT, code 9) instructs a previously addressed device to take over as controller
      of the bus.

The UCG commands are issued to all devices which are present on the bus (although that does not mean they can take the rquired action) and are:

Local Lockout (LLO, code 17) causes all devices which are capable of doing so to disable
      their front panel controls (so that they are under the control of the bus without fear of
      interference).
Device Clear (DCL, code 20) causes all devices to clear.
Parallel Poll Unconfigure (PPU, code 21) prevents all devices from responding to a parallel
      poll (issued so that new configure commands can be given).
Serial Poll Enable (SPE, code 24) informs devices on the bus that a serial poll follows.
Serial Poll Disable (SPD, code 25) informs them that it is finished.

Table 8.2   The nmemonic codes used to indicate the
interface functions supported by a GPIB device

nmemonic	function name
T,TE	Talker or Extended Talker
L,LE	Listener or Extended Listener
SH	Source handshake
AH	Acceptor handshake
RL	Remote/Local
SR	Service Request
PP	Parallel Poll
DC	Device Clear
DT	Device Trigger
C	Controller
E	Driver type

While the majority of data transfers can take place under the control of a BASIC program, it is important to realise that not all micros have bus commands implemented via BASIC. Thus on several micros (and some GPIB interfaces) bytes with bits 5 and 6 as 0s cannot be output directly from BASIC while the ATN line is held lo. Of course it is also important to appreciate that devices labelled as GPIB compatible will probably incorporate only a limited number of GPIB functions. With the revision of the IEEE 488 standard in 1978 came the recommendation that all GPIB devices should be marked near the GPIB connector with a set of codes to indicate which GPIB interface functions the unit supported. A list of the available mnemonic codes is given in table 8.2, although in use these are all followed by numeric codes which indicate something about the implementation. (A 0 means not implemented, 1 means full implementation, and other values mean partial implementation. Naturally there are exceptions - E1 means open collector bus drivers, while E2 means TRI-STATE drivers.) Extended Talkers and Extended Listeners are devices which can use two bytes for addressing as Talkers or Listeners, the second byte being transferred as a secondary address code.

8.4 Examples of GPIB adaptor cards for the PC

There are numerous IEEE 488 interface adaptor cards available for the PC and these allow the PC to be connected to a GPIB and control data transfers over the bus. While the handling of data, control and management busses can be performed at the assembler level, either by writing the appropriate code or by using routines of the computer's operating system (if available), it is generally convenient to communicate with GPIB devices using an applications program or a high level language such as BASIC. Many of the commercial GPIB devices transmit and receive data in the form of ASCII character strings, so BASIC is convenient for handling communications. Relatively simple BASIC-type commands are available with most GPIB cards supplied for PCs, although different approaches have been adopted by different manufacturers. While it is not possible to cover many of the commercially available cards in a book of this kind, two examples which illustrate the different approaches are discussed briefly below. These examples have been chosen because they represent the two major approaches adopted by card manufacturers to simplify bus handling from BASIC programs, and because we have used these specific cards in our laboratories for some time.

8.4.1 The CIL GPIB interface card

The CIL GPIB interface is a full length card containing its own Z80 processor, either 8 or 40 kbytes of RAM (options), and a software support package in EPROM. The interface may operate independently of the host computer, but at the same time has a relatively simple communication interface. Installing the card is straightforward, although some adjustment of the on-board switches to set IO addresses may be required if the host PC contains many other cards. The card uses IO addresses of 0300H - 0307H, although these may be changed by switches to start at 0400, 0500,0600, or 0700H, and memory addresses C000H - C200H, which may also be changed to start at D000 or E000H. On the versions which I have used a standard, stackable IEEE connector fits with its cable pointing down to the laboratory bench. This is unfortunate as IEEE cables can be relatively difficult to bend to a small radius curve, and to avoid this I have preferred to use ribbon cable connections between the interface card and the first IEEE device on the bus.

As the card contains all the software for actually handling communication over the GPIB, user's programs are merely required to issue CALLs to on-card routines and provide values or variable names for parameters such as device addresses, data for output, destination for input and status strings, and error codes. Calls are made through a jump table, so the user's program needs to define the absolute address of the start of the jump table, then call the required routine with an offset address. For example, if the card is installed with memory addresses of C000H - C200H, then the jump table starts at C000H. A BASIC program can CALL offsets from this address by having the command

        DEF SEG &HC000

somewhere in the program before the first call. Offset values may be stored in variables with names which are helpful in suggesting the function, and the manual recommends the following (among others):

            DATAOUT = 0
            DATAIN = 3
            STATUS = 6
            CMD   = 9

A CALL to DATAOUT is then simply

            CALL DATAOUT (AD%, DA$, ST%)

where AD% is the address of the IEEE device which should listen to the data, DA$ is the data string being sent, and ST% is a variable which contains the status code on return to BASIC.

A variety of commands may be sent to the CIL card using the BASIC CALL

CALL CMD (ER%, Command$, ST%)

where ER% returns with a code which, if non-zero, represents an error (such as no device present), Command$ is a command string, and ST% will contain a status code. The command string may be used to define a secondary address, issue an IEEE bus command (GET, GO TO LOCAL, DEVICE CLEAR, etc), a discrete code (eg MTAn, MLAn), operate IFC or REN, carry out a serial poll, or define string terminators for data input or output. In fact the commands provide virtually all the facilities required to control devices which adhere to the IEEE specifications. [Not unreasonably, they do not permit complete communication with slightly non-standard devices, such as the old Commodore PET peripherals, as these require command bytes to be transmitted with bit 7 of the data bus operational.]

This approach to the PC IEEE interfacing problem has a number of attractions. It is relatively straightforward and even a novice BASIC programmer can have a system working in a few hours; the on-card memory allows data transfers to be buffered, so that data readings (or output) need not be delayed by a slow user program; and the on-card software means that negligible computer memory is lost to interface communications. On the other hand, this approach leads to some difficulties through the subroutine calling conventions of different languages; for example, C and Pascal expect subroutine calls to handle parameter passing in different ways. To use the technique outlined above with QuickBASIC or Bascom requires that the CALL commands are modified so that

CALL DATAOUT (AD%, Da$, ST%) in interpreted BASIC becomes

CALL Absolute (AD%, DA$, ST%, DATAOUT) in QuickBASIC.

8.4.2 Brain Boxes PC - IEEE 488 interface card

The Brain Boxes IEEE 488 interface comes as a half length card, complete with an instruction booklet and a disk of software. The default IO addresses are 0300H and 0301H, however, it is a simple matter to reset the IEEE card to other addresses (eg. 03E0H) if the existing addresses clash with those of other cards within the PC. The standard IEEE connector protrudes from the rear of the card space with the cable going upwards. The Brain Boxes IEEE interface is based on the Motorola 68B21 peripheral interface adaptor chip (and IEEE transceivers). This results in the card containing a relatively small number of ICs compared with those interfaces built around microprocessors and ROMs. On the other hand it also means that the software to control bus activity must reside in the PC. The instruction manual recommends that users do not attempt to write their own software for operating the bus, but use instead the two alternatives provided: an MS-DOS device driver, which allows the operating system to treat the interface card like a file, or a number of assembler object code modules (complete with a jump table) which can be linked into user's programs. In spite of these provisions a memory map of the PIA's 6 registers is included in the manual and this provides all the details necessary for controlling the individual lines of the IEEE bus.

The device driver for MS-DOS is provided on the disk accompanying the IEEE card. It is installed in the same way as any other device driver, with a command of the form

DEVICE = IEEE.SYS

included in the CONFIG.SYS file of an MS-DOS disk. An example CONFIG.SYS is included on the software disk and may be copied directly to a user's disk. Once installed the driver may be used in a variety of ways, including redirecting program input and output (with <IEEE and/or >IEEE), by COPYing data between the IEEE port and a data file or other port (such as LPT1:), or by issuing the DOS PRINT command and responding to the [PRN]: prompt with IEEE. Alternatively the driver may be installed with redirection instructions. For example:

DEVICE = IEEE.SYS /"COM1   ",4,7

redirects data transfers specified for COM1 to the IEEE port, using primary address 4 and secondary address 7. This allows very straightforward IO operations from users' programs which would be written as though they were accessing the RS232 serial port. (Naturally COM1: is not available for normal use while its transfers are redirected.) For programs which will control only a listening device (such as a printer or plotter) the output of LPT1: may be redirected in the same manner.

The IEEE bus can be used directly from assembly language by using MS-DOS interrupt 21H and passing codes to specify required functions (such as open, close, read, write, or the IOCTL function - which exchanges bus control or status information with the device driver). A complete working assembly language program is included on the software disk (it is called MSTEST.DOC) and may be compiled using Microsoft's Macro Assembler. This program provides a useful collection of examples of handling the IEEE bus from assembler - including the ability to send and receive the full range of IEEE instrumentation commands and automatic SRQ action.

The IEEE device driver also allows IEEE activity to be controlled by BASIC programs using the standard basic IOCTL statement and IOCTL$ function. In this case the user's program needs to open two files with:

OPEN "IEEE" FOR OUTPUT AS #1
OPEN "IEEEDATA" FOR OUTPUT AS #2

(yes, they allow input, but the book says open them as shown). From then on all communication may be handled by commands such as:

100 IOCTL#1, "4,7"

which specifies that the current IEEE device is device 4 with a secondary address of 7,

       200 A$ = IOCTL$(1)

which returns in A$ a string containing the current primary and secondary address and a host of other status information,

       300 IOCTL#2, DA$

which sends the string DA$ to the current IEEE device, and

       400 DI$ = IOCTL$(2)

which reads in a data string from the current IEEE device - terminated by EOI (unless the user has specified an alternative terminator).

This approach to IEEE communications compares favourably with CALLs to on-card ROM routines as used in other cards. It is very straightforward and may be incorporated into BASIC programs ported over from other micros such as the Commodore PET without difficulty. Unlike the ROM CALLS approach, using the file technique permits strings to be input without the string space needing to be defined (and space filled) in advance. Furthermore, the Brain Boxes technique is fully compatible with Microsoft's QuickBASIC, C, Pascal, Fortran, and indeed any language that runs under DOS 2 and above.

In addition to the transfer of strings over the IEEE bus, a large number of commands are provided for controlling the bus. All are sent to the device driver using IOCTL#1 and include ATN (to lower ATN), IFC (to pulse IFC), LOCAL (set device to local), REMOTE (set remote enable), TRIGGER (perform GET), SPOLL (perform a serial poll), and WAIT for a variety of things.

The old Commodore PET microcomputer was provided with a somewhat non-standard IEEE 488 interface, and indeed used this interface for the connection of disk drives and other peripherals to the computer. One particularly valuable feature of the Brain Boxes IEEE card is its ability to communicate with Commodore IEEE devices in Commodore's way. In fact, with the example programs provided it is a relatively simple matter for the user to write his own routines in BASIC to handle file transfers between the PC and a Commodore disk drive on the IEEE bus. However, the software disk includes a program which handles transfers of sequential files from the Commodore drive to the PC. The program provides the user with the options of adding linefeeds to all carriage return characters, performing PET - ASCII conversions, or ignoring bit 7 of bytes transferred.

CHAPTER 9

THE SERIAL STANDARD INTERFACE AND COMMUNICATION SYSTEM

9.1 Serial standards

Entirely different standard systems, primarily intended for data transfer rather than instrumental control, are used for connecting computers to the telephone system and one of these has become popular for the connection of peripherals, such as microcomputers or printers, to laboratory instruments. This is the RS232 standard of the Electronic Industry Association, which has undergone two revisions and in its present form is called the RS232C standard. This often provides a convenient means of connecting a microcomputer to a commercial instrument equipped with an RS232C interface, particularly as many computers (such as the Amiga, Atari, Macintosh and many PCs) are now fitted with RS232C-compatible ports, or have RS232C interfaces available as extras. In fact it is intended that the RS232C standard will gradually be replaced by the RS449 and RS422 standards which make allowance for the electrical characteristics of modern integrated circuits and permit greater operating distances and data transfer speeds. However, it is also intended that equipment conforming to the newer standards can interoperate with equipment designed to RS232C standard, and yet another standard, the RS423, exists to provide a high degree of compatibility with RS232 systems while conforming to the electrical and mechanical specifications of the RS449 and RS422 standards. As the overwhelming majority of the serial interfaces on microcomputer and laboratory equipment currently available are claimed to be RS232C standard, this discussion will be concentrated on that standard.

9.2 Serial data transfers

The particular attraction of the RS232 system is that it is a serial data transfer system. Rather than having eight wires which each carry the signal level corresponding to one bit of a byte (the parallel system), a serial system can use as few as two wires (a signal wire plus ground) to carry the signal level of one bit at a time. This is probably the most valuable feature of the system in that relatively widely spaced units may be connected with minimum amounts of cabling. The standard specifies allowed speeds (up to 20 kilobits per second) and signal levels of data transfers (between ± 3 V and ± 25 V, no TTL or CMOS logic levels here), which take place in bit serial mode over distances of up to 15m. Although the standard does not specify a particular character code for the transmission of characters, most modern RS232C systems of interest here use the asynchronous serial data format, illustrated in fig 9.1, to transmit a single character of data along the wire (it need not be 8 bits, so we use the

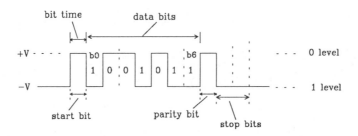

Fig 9.1 *The asynchronous serial data format normally associated with RS232C*
systems. The example shown is the character code 105 (the ASCII character
i), even parity with two stop bits. Note that a -V level represents logical 1,
the first data bit is the LSB, and the line idles at the 1 level.

word character rather than byte). (The RS232C standard also covers synchronous
transmissions in which a clock signal is used, but we shall not be concerned with this mode
as it is rarely used in microcomputer systems.)

The transmitted information consists of a series of signals, each at one of two allowed
levels - the 0 level or the 1 level (which are usually voltages such as +6 V for level 0 (the
space level) and -6 V for level 1 (the mark level)) - and lasting for a specified period of time.
Before the sequence starts the transmission line is held at level 1, then a "start bit" of 0 level
is sent to indicate that data will follow. The start bit is followed by up to eight timed
applications of level 0 or 1 making up the data bits as shown in fig 9.1, and then one (or two)
"stop bits"of level 1 to indicate the end of the data and ensure that the line is held at level 1
before the next start bit is transmitted.

Clearly the timing of the signals is very important for the transmission of a series of bits
in this manner, and the rate of bit transmission within a series of bits is called the Baud rate
(eg 300 Baud means 300 bits per second - although they are not all data bits). The most
commonly used Baud rates are 110 (the old standard mechanical Teletype rate), 300, 1200
and 9600. It should be noted that each 8 bit character transmitted actually requires 10 or 11
bits to be sent, so that operating at 300 Baud using 10-bit format, the maximum character
transfer rate is 30 per second. Furthermore, the transmission is asynchronous in that start and
stop bits are used to signify the beginning and end of each character. The transmission of the
individual bits of a character does depend on controlled short-term timing, but this timing is
important only for the duration of each character, so high precision is not required. One of the
advantages of asynchronous transmission is that characters need not be contiguous in time,
but may be transmitted as they become available (as for example, from a manually operated
keyboard).

Most modern RS232C systems of interest here use 8 data bits to transmit a single
character, although 5, 6 and 7 data bit systems may also be encountered, and, while any
combination of bits may be used, the most common code for data transmission is the
American Standard Code for Information Interchange (ASCII) shown in appendix 2. The
code utilises only seven bits, the 128 characters available covering upper and lower case

letters, digits, punctuation marks and about 30 control characters such as <carriage return>, <line feed> and <escape>. The eighth bit is sometimes sent as a 0, but may also be used for parity checking (which helps to ensure that the other seven bits were correctly received). Other codes are found on RS232C systems, including Baudot and Murray codes (5 bit), IBM Correspondence Code (6 bit) and EBCDIC (8 bit). Computers which are already fitted with an RS232C port are usually able to transmit or receive data using standard BASIC IO commands such as INPUT# and PRINT#, and major operating systems such as MS-DOS, Unix and AmigaDOS allow RS232 port IO to be treated as any other file IO. However, there are a number of variables associated with any RS232C interface, and while most RS232C interfaces for microcomputers allow the selection of all relevant parameters, some may be very restrictive. It is essential to check any microcomputer interface before purchase to determine what selections may be made of:

1. Allowed signal levels, ±3 V to ±25 V. Output is generally not a problem, as anything above +/- 5 V should be suitable. However, the maximum allowed levels for input can cause difficulties. The interface should be able to receive ±25 V without damage.
2. Baud rate, typically covering the range 110-9600 Baud, although much broadcast traffic occurs at 40/50 Baud. The "standard" rates are: 50, 75, 110, 134.5, 150, 300, 600, 1200, 1800, 2400, 3600, 4800, 9600 and 19200).
3. Number of data bits, normally 5, 6, 7 or 8 bits per character.
4. Number of stop bits, should be selectable as 1, 1.5 or 2. (1.5 bits means that the stop bit has a duration 1.5 times that of the data bits. 1.5 stop bits are rarely encountered these days.)
5. Parity check or ignore. Both options should be available.
6. Parity even or odd. Again, both options should be available.

It is very important to bear in mind that not all serial equipment uses the RS232C standard signal levels, even though much else may be common. There are in particular two important serial systems which are still in widespread use but which can only be connected to RS232C systems with the aid of a level-changing interface. The first is the (US) MIL standard 188C which differs from RS232C in having the signal polarities reversed. Generally connecting a 188C device to an RS232C device will not result in damage to either, although they won't talk to each other. A more serious problem arise with the variations of high level interfaces, frequently encountered on old Teletypes (such as the Model 33) and teleprinter equipment. These interfaces operated with fairly high voltage levels (typically 60 - 120 V) and currents adequate to drive mechanical solenoids (20 - 60 mA). Furthermore, quite a lot of modern serial equipment uses signal levels compatible with 20 mA current loops (known as 20 mA loop), although at lower voltages. Connecting these systems to RS232C devices can result in damage to the RS232C circuits. Level changing with a single transistor or opto-isolator is quite straightforward.

The advantages of the RS232C standard interfaces lie primarily in the ability to use relatively long lengths of connecting cable, which may consist of as little as two wires for one-way transfer or three wires for both-way communication, and the relative ease with which signal levels on such wires may be changed (eg from TTL 0/5 V to ±12 V , or ±12 V to 20 mA loop using simple IC transceivers). In principle the lower voltage level implementations of the RS232C systems are intended for communication distances of less than 15m, although thousands of metres can be handled with the 20mA loop system. However, in the laboratory the limitation of only one device per computer serial port can be severe. Serial data transfer can be rather slow even though it may be theoretically possible to operate at high Baud rates, and this is particularly true if the system is being driven from BASIC or if the software does not allow the operation of any kind of handshake. Nevertheless the existence of an RS232 connection on many larger instruments such as liquid scintillation counters, multichannel analysers and X-ray diffractometers, does make the coupling of a computer relatively simple and, in many cases, a microcomputer and printer may be cheaper than a conventional terminal or Teletype.

In passing we may note that the newer standards do offer dramatically improved data transfer speeds. Thus the RS422 standard permits a maximum transmission rate of 10 Mbits per second, while the RS423 (which uses levels of ± 5 V) permits up to 100 kbits per second transfer rates. They also specify cable lengths up to 60m, although allowing "tailored" operation over very much greater distances.

9.3 The RS232C connection

The RS232C standard (which runs to 29 pages) begins with the words: "This standard is applicable to the interconnection of data terminal equipment (DTE) and data communications equipment (DCE) employing serial binary data interchange". Thus fundamental to the philosophy of the RS232 system is its asymmetry, one device being regarded as a DCE (such as a modem - a modulator/demodulator which changes the signal levels into tones for transmission over the telephone network) and the other as a DTE (ie a terminal). The arrangement envisaged in the standard is illustrated in fig 9.2, where it is important to note that the RS232C standard refers to the connection between the terminals and their corresponding modems; it does not refer to the connection between the modems, which is covered by an entirely different range of standards specified by the Bell standards or the V and X standards of the CCITT (Comite Consultatif Internationale de Telegraphique et Telephonique - part of the United Nations agency the International Telecommunication Union). The RS232C standard specifies that a cable ending with 25 pin "D" connectors is used for the DTE-DCE link, the DCE itself having a socket (female connector) and the DTE a plug (male connector).

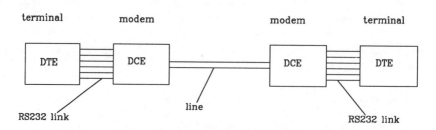

Fig 9.2　　*A complete communications link employing RS232C standard connections between each terminal and its associated modem.*

pin no.	direction	abbreviation	name
1	-		protective ground
2	to DCE	TxD	transmitted data
3	to DTE	RxD	received data
4	to DCE	RTS	request-to-send
5	to DTE	CTS	clear-to-send
6	to DTE	DSR	data set ready
7	-	gnd	logic ground
8	to DTE	DCD	carrier detected
9			reserved
10			reserved
11			unassigned
12	to DTE		secondary carrier detected
13	to DTE		secondary clear-to-send
14	to DCE		secondary transmitted data
15	to DTE		transmit clock for synchronous data
16	to DTE		secondary received data
17	to DTE		receiver clock for synchronous data
18			unassigned
19	to DCE		secondary request-to-send
20	to DCE	DTR	data terminal ready
21	to DTE		signal quality detected
22	to DTE	RI	ring detected
23	to DCE		data rate selector
24	to DCE		transmit clock
25			unassigned

Fig 9.3   *The pin connections of the 25-way connector used for RS232C systems.*
(Rear view of socket.) Note that some printers now use pin 11 as a
handshaking line to indicate "ready for data".

The full RS232C link consists of two channels, a primary and a secondary, both of which are able to transmit or receive, although it was always envisaged that the primary channel would be the one on which the data would pass while the secondary channel would be reserved for small amounts of control information. The names of the signals and the connector pin numbers are given in fig 9.3. Note that separate lines are used for transmitted data and received data - the data lines between the DTE and DCE are not bidirectional. Just to ensure that we all understand the system used in the RS232C standard note that the data signals (on pins 2 and 3, and 14 and 16) use negative logic, eg. -12 V is a 1, +12 V is a 0, while the control signals are all positive logic, eg. +12 V on pin 5 tells the DTE that CTS is true and the circuit is clear for it to send data.

9.4 Data transfers

For our present purposes we may consider data transfers within a restricted system consisting of a DTE and DCE alone. This description will ignore a number of control lines which are used in communication systems which resemble fig 9.2, and we will return to these in section 9.7. To reduce the jargon content of our discussion of these systems we will refer to the DTE as a terminal and the DCE as a modem for the remainder of this chapter, and hope that this will serve to clarify the principles of handling RS232C data transfers. However, it is important to bear in mind that in this discussion "terminal" really means any RS232C device wired as a DTE, and "modem" really means any RS232C device wired as a DCE. An RS232C terminal and modem are able to communicate with one another only when the terminal finds that the DSR line is true (high) and the modem finds that DTR is true - ie. both ends of the RS232C link find that the other end is ready. The remaining data transfer activities fall into three stages:

1. The terminal and modem agree that data transfer may occur.
2. Data transfer takes place.
3. The terminal and modem agree that data transfer is at an end.

Stage 1 involves three possibilities: either the allowed data transfer will be from the terminal to the modem, or it will be from the modem to the terminal, or it may be in both directions (remember, there are two data lines, transmitted data and received data, so it is quite legitimate to have data transfer in both directions simultaneously - provided that the terminal and the modem can handle the signals). Data transfer in only one direction is referred to as Simplex communication if the one direction is fixed and unalterable, or Half-duplex (HDX) if the direction may be changed. If the data transfer may be bidirectional then the communication is called Full-duplex (FDX). For simplex communication there are no alternatives once the terminal and the modem are ready, so simplex systems generally ignore all other control lines and the equipment supplying the data just gets on with the job. A simplex system really needs only the correctly positioned data line and ground. (DTR and DSR may be used, but could equally be hard wired to high levels to economise on cabling.) On the other hand, simplex communication is relatively inflexible, requiring the receiving device to be capable of receiving data at any time. Thus if a printer were to be operated in simplex mode it would have no option but to carry on receiving data long after the paper or ribbon had run out. Simplex systems are rarely encountered these days, although they are worth remembering for simple in-house serial interfaces dedicated to specific functions.

Half-duplex communication requires that the device wishing to send data signals the other device so that both can agree on the direction of data transfer. When the terminal wishes to send data to the modem it (the terminal) first raises the RTS (request to send) line to indicate that it wishes to send data to the modem. The modem checks the DCD (carrier detected) line to ensure that it is false (ie to ensure that the modem is not trying to send data to the terminal), then if the modem is happy with this arrangement it (the modem) replies by raising the CTS (clear to send) line to true, acknowledging that it is able to receive data. If the terminal sees CTS go true it knows that data transfer may occur, and it maintains RTS true until it (the terminal) has finished sending data. If the terminal does not see CTS go true it knows that transfer may not proceed, and it (the terminal) drops its RTS line and starts all over again. When the modem wishes to send data to the terminal it raises the DCD line to true, preventing the terminal from taking control of the line as above. Thus the direction of data flow is determined by which device raises its control line first, either the terminal raising RTS or the modem raising DCD. In either event one end of the link is giving the other end of the link full control over the link. Clearly half-duplex communication offers improved versatility over simplex mode, in that data transfer in either direction may take place.

In full-duplex communication RTS, CTS and DCD are held true constantly, unless a major failure occurs in which case the modem drops CTS and the terminal treats this as an error condition. The terminal may send data to the modem and the modem may send data to the terminal at any time.

Stage 2 (the data transfer) may now take place. Actually the RS232C standard does not permit the control lines to be used to handshake individual characters of data - once the terminal sees CTS go high, it expects to be able to transmit as much data as it wants, and the modem is not permitted to interrupt by, for example, dropping CTS (except to indicate an error condition - in which case the transfer is regarded as terminated). Similarly if DTR is true and a modem is transmitting to the terminal, it does not expect to be interrupted by, for example, DTR going false. In spite of this a number of manufacturers have produced serial devices, such as printers, which do have character handshaking implemented using a pair of control lines. Unfortunately different manufacturers have chosen different pairs of control lines, some using the DTR/DSR pair, others using RTS/CTS, DSR/CTS or DTR/RTS, and others using pins specified in the standard as being unassigned. Such an independent approach towards the implementation of standards does lead to a certain amount of confusion amongst users attempting to connect one piece of equipment to another, and is at least in part responsible for the relatively poor image of serial interface systems. To accommodate such systems it may actually prove necessary to produce interconnecting lead with some surprisingly individual wirings, such as connecting DTR to CTS and RTS to DSR! Systems which implement non-standard features of this kind are best avoided because of the constraints they place on the user's freedom to change computers or languages.

Some kind of data handshaking is virtually essential when using devices of limited memory size or other features which could require a temporary cessation of data flow (eg. the printer running out of paper). Fortunately both half-duplex and full-duplex communication channels provide an obvious means of implementing a degree of handshaking - in that they permit the receiving device to send messages back to the transmitting device. The precise mechanism adopted for handshaking is referred to as a protocol. It is relatively unusual to be able to choose a protocol unless both devices are programmable; usually one device will have a single built-in protocol, and it will be up to the other device (generally the computer) to be able to understand and operate that protocol. While many protocols for data transfer exist, the most commonly encountered fall into one of the following categories: hardware XON/XOFF, software XON/XOFF, and Enquire/Acknowledge.

XON/OFF is a data transfer protocol in which the device receiving the stream of data sends control characters to the device transmitting the data to tell it to start or stop the flow of

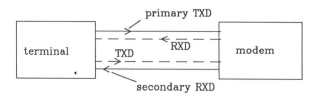

Fig 9.4   *A half-duplex RS232C link in which the primary channel handles*
           the data characters (left to right) and the secondary channel
           handles the XON/XOFF control characters (right to left). (Control
           signal lines not shown.)

data. Generally the ASCII character DC1 (device control 1) is used to start the flow of data (XON) and DC3 is used to halt the flow (XOFF). However, other characters may be used for XON and XOFF, and some software allows the user to specify which characters are to be used in these roles. In a half duplex communication link (in which information may flow in only one direction at a time) the devices must make use of the RS232C secondary channel (see fig 9.4) to send the XON and XOFF characters, so, naturally, the wires carrying the secondary channel data and control signals must be present in the link. Because the XON/XOFF control is exercised through the RS232C's secondary channel wiring, this protocol is called Hardware XON/XOFF. In full-duplex systems it becomes possible to transmit the XON/XOFF codes via the primary channel connections, even while data is being received over the primary channel. This is called Software XON/XOFF, and may be implemented over a link in which the secondary channel is not physically connected. In any event, the important characteristic of the XON/XOFF approach is that the same device (the data receiver) is responsible for sending out both control codes and the other device must be able to respond accordingly. XON/XOFF control is particularly popular amongst the manufacturers of devices which are thought of as one-way devices, such as printers.

An alternative approach requires the devices at either end of the serial link to be able both to send control codes and to respond to received control codes. Typically a data transmitting device will send out its data in blocks of, say, 200 characters each. Each block will be separated from the next by the transmission of an Enquire character (ASCII ENQ) and the transmitting device will wait until it receives an acknowledge character (ASCII ACK) before it continues transmitting. This Enquire/Acknowledge protocol is found on a number of buffered peripherals, such as the Hewlett Packard graphics plotters, where the acknowledge signal is withheld by the receiving device until there is enough room in the buffer for another complete block of data. Again, in principle, the RS232C secondary channel could be used for the acknowledge characters, although an advantage of the ENQ/ACK protocol is that, as each device can know when it needs to transmit and receive, the technique can be used on the primary channel of either a half-duplex or full-duplex system.

Stage 3 (agreeing on end of transfer) applies only to half-duplex systems, as full-duplex systems maintain fixed control levels as long as both terminal and modem are operational - so, in a sense, the RS232C system is never informed that data transfer is at an end until either or both devices are switched off. In a half duplex system the transmitting device signals that it has finished transmitting by dropping to false its RTS line (for a terminal) or its DCD line (for a modem). If it is a terminal that finishes transmitting, then the attached modem responds by dropping the CTS line. In each case the system has reverted to an idle state in which only DTR and DSR remain true, and either the terminal or the modem may initiate fresh activity as in stage 1.

9.5 Connecting RS232C systems

RS232C connections on microcomputers and other devices may be wired as terminals or as modems, and, as discussed above, these terms really mean any device wired as a DTE or DCE respectively. The difference between a DCE and DTE is not always as obvious as, say, the difference between a transmitter and a receiver using a Centronics interface, although it may be helpful to remember that the RS232C control lines are named from the point of view of the DTE. Of course the real difficulty is that both the DTE and DCE can transmit and receive characters. The difference between them lies in the use of the many other lines of the 25-wire connection. For example, the DTE can control the levels on the RTS and DTR lines, while the DCE controls CTS, DSR, DCD and RI. Unlike GPIB systems, many of the commonly encountered RS232C systems (other than real modems) utilise very few of the available signal lines, so difficulties can arise when two (nominally) RS232C systems which use different numbers of control lines are connected together.

To establish communication between two items of RS232C equipment we need to know three things about the link:

1. How each item is wired up (ie as a modem or as a terminal) - so that the two can be correctly connected together.
2. What characteristics are needed for the serial data (ie. Baud rate, number of bits, parity, etc.) - so that the two ends will understand one another's characters, and
3. What protocol is to be used for controlling the data transfers - so that successful communication can take place.

We will examine each of these properties of the link in turn.

In principle a modem-type device transmits data on pin 3 of its socket, and receives data on pin 2, while a terminal transmits on pin 2 and receives on pin 3 of its plug. In most cases commercial equipment with RS232C connections is hardwired and no choice is left to the user. Furthermore most RS232C implementations on microcomputers and their peripherals are wired up as terminals. The principal difficulty is that some manufacturers do not conform to the RS232C specifications, in some cases preferring to use a socket connector on a microcomputer, printer or plotter which will operate as a terminal - presumably to prevent metal or fingers from coming into contact with the pins which would be present otherwise. Others prefer to produce their equipment to operate as a modem, in spite of the fact that the equipment may bear no relation to the modem function indicated in fig 9.2 - presumably to simplify the cabling required. A more serious problem arises where manufacturers produce equipment on which the "spare" pins of an RS232C connector are used for completely

Table 9.1 The 25-pin serial port of the Amiga

Pin	function	Pin	function
1	ground	14	-5 V
2	TXD	15	audio out
3	RXD	16	audio in
4	RTS	17	716 kHz
5	CTS	18	interrupt 2
6	DSR	19	
7	ground	20	DTR
8	DCD	21	+5 V
9		22	
10		23	+12 V
11		24	3.58 MHz
12		25	reset b
13			

different purposes. For example, the serial port of the Amiga is detailed in Table 9.1. Connecting this to another item of equipment on which pins such as 21 or 23 are grounded can lead to blown fuses, or worse.

In short, it is not always certain that the role of a specific item of equipment stated to conform to RS232C specifications will be correctly identified by an examination of its connector. Indeed it may not be wise just to plug in and hope for the best. A better approach is to attempt to identify which pins on the connector are operating as outputs - ie is it pin 2 or pin 3 which is idling with a negative voltage on it. For this purpose a device known as a break-out box is particularly useful, as these generally contain indicator lights which identify which lines are acting as outputs when an RS232C device is connected to it. Of course the best approach is to look in the device's manual and be quite certain about which pins do what, and then to ensure that only the connections actually required for communication in a given system are provided by the cable.

Once the identification of the items to be connected is clear it becomes possible to identify three types of connection: a terminal connected to a modem, a terminal connected to a terminal, or a modem connected to a modem. The more straighforward connection involves a terminal connected to a modem. In this case all the pins on the terminal's male connection may be connected to all the corresponding points on the modem's female connector using a cable referred to as a "one-to-one" cable. In practice it is unusual to need all 25 wires, particularly as most microcomputer serial connections are now full-duplex - so that only the primary channel is required. Fig 9.5 illustrates a typical terminal to modem connection using a one-to-one cable wired for the primary channel only. The extension to both channels should be obvious.

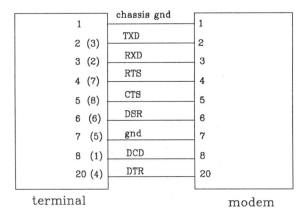

Fig 9.5   *Typical single channel DTE-DCE connection using a one-to-one cable.*
           (Figures in brackets indicate pin numbers for the 9-pin connectors found on the
           PC/AT. Note that 9-pin implementations do not include a protective ground)

Where two devices operating as terminals are to be connected it is necesary to use a
connecting cable which essentially fools each terminal into believing that it is actually
connected to a modem. A cable of this type allows the terminal to behave as if it were part of
the circuit shown in fig 9.2 even though the two modems are actually missing, and for this
reason is referred to as a "null modem" cable. In fact there are many possible ways of wiring
up a null modem cable. Fig 9.6 illustrates one primary channel null modem connection, and
it will be seen that transmitted data from one terminal is passed to received data at the other,
DTR on one is passed to DSR on the other, and RTS on one is passed to its own CTS pin and
to DCD on the other. The signal reference ground and the protective grounds must of course
remain connected on a one-to-one basis. As most microcomputer operating systems expect
their serial ports to be configured as terminals, this terminal-terminal connection is probably

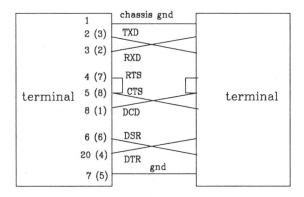

Fig 9.6   *Typical single channel DTE-DTE connection using a null modem cable. (Figures*
           in brackets indicate pin numbers for the 9 pin connectors found on the PC/AT.
           Note that 9-pin implementations do not include a protective ground.)

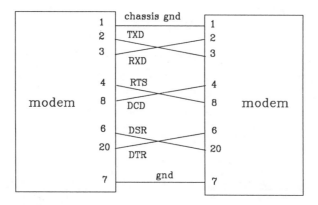

Fig 9.7   *A possible arrangement for a single channel DCE-DCE connection.*

the most commonly encountered RS232C arrangement. It also results in a relatively straightforward programming of data interchange, allowing, for example, easy copy of files between two PCs using the MS-DOS copy command (see below).

It is possible to connect together two devices operating as modems, although with most modern microcomputers being configured as terminals it is rather rare for such connections to be required. In my experience connections of this sort are only required when an equipment manufacturer has taken an unusual view of his own equipment. An example of the cabling required for such connections is illustrated in fig 9.7.

Figures 9.5 - 9.7 might be thought to encompass all reasonable possibilities. Unfortunately the variety of free thinking which has been adopted by manufacturers of devices provided with nominally RS232C connections has been such that there are many others. In Martin Seyer's book "RS232 Made Easy," the author details forty- two different interconnection cables for connecting computers to terminals and printers. One of the hidden benefits of the IBM PC's success is that it has forced some sense of order into the minds of many peripheral manufacturers, so that gradually the serial interface arena is becoming less crowded with connections calling for different cables.

9.6 The PC serial port

The PC clones may be fitted with asynchronous serial communication ports on plug-in cards, although such ports are often incorporated on other cards, such as dispay adaptors. Most follow the IBM practice of implementing a primary channel subset of the RS232C standard, with the port configured as a terminal. With the original PC and the PC/XT IBM adopted the normal 25-pin D male connector protruding from the rear of the machine. However, with the introduction of the AT, IBM changed to the smaller 9-pin D male connector, and many clone manufacturers have adopted this 9-pin "standard" for micros based on 8088, 8086 and other processors. The pin arrangement for both connectors is summarised in table 9.2. The typical one-to-one and null modem cabling pin connections are given in the brackets within figs 9.5 and 9.6.

Table 9.2 The pin configuration of the PC serial ports

25-pin D connector	signal	9-pin D connector
2	TxD	3
3	RxD	2
4	RTS	7
5	CTS	8
6	DSR	6
7	ground	5
8	DCD	1
20	DTR	4
22	RI	9

The PC's operating system contains facilities for controlling two serial ports simultaneously, and these are referred to as COM1: and COM2:. They may be operated as full- or half-duplex channels and may have their data parameters set by direct user command to the operating system or by application programs through a call to the ROM BIOS through INT 14H. In high level languages this is generally accomplished by OPENing a file named COM1 (or COM2) with parameters to specify the data charateristics. Thus to set up a serial port to send data to an external device using 9600 Baud, 8 data bits, 1 stop bit and no parity, one could issue the DOS command

MODE COM1:96,N,8,1

or, from BASIC, the command

OPEN "COM1:9600,N,8,1" AS #4

Note that the MS-DOS MODE command allows the Baud rate to be specified using at least the first two digits of the required rate - thus the value could have been specified using 96, 960 or 9600.

Data transfers may also be carried out using direct commands to the operating system or via calls to the ROM BIOS (INT 14H) or MS-DOS (INT 21H). Again high level languages generally handle serial IO via the normal filing system calls, such as INPUT# and PRINT# in BASIC. Thus to send data to an external device after the data characteristics have been set up as described, one could use the DOS command

COPY filename COM1:

or BASIC commands of the form

PRINT#4, X$

While it may sound as though the operating system, high level languages and applications all handle all aspects of communication via the serial port, unfortunately this is not quite the case. MS-DOS in fact incorporates rather little support for the serial port. It cannot, for example, check the status of the port's control lines - it leaves this job to the user's program. Furthermore, the MS-DOS serial "device" is unbuffered and is not interrupt driven. Consequently if a stream of characters is sent to the serial port from an external device at a speed greater than an MS-DOS application can process them, then some characters may easily be lost although no error is reported. [The stdaux stream provided in Microsoft C is also unbuffered.] In practice most applications which need to use a serial port do so with their own software to handle data transfers. Thus BASIC provides buffered serial input and output, and provides the user with the ability to operate either with or without the RS232C control lines RTS, CTS, DSR and DCD. These facilities are provided through the OPEN command in the form of additional parameters which may specify the suppression of the RTS signal or the time limits for a wait for a response on the incoming control lines CTS, DSR and DCD - with a time limit of zero causing the lines to be ignored. For example

OPEN "COM1:9600,N,8,1,RS,CS0,DS0" AS #4

opens the serial port as above, but without RTS being used and with any signals on CTS and DSR being ignored. [DCD is ignored by default.] This allows a bidirectional link to be established with just the three wires needed for transmitted data, received data and ground - a connection which is much more difficult to achieve through MS-DOS commands alone. It turns out that BASIC is rather well suited to handling serial data transfer through the presence of its support for the COMn channels. Of course, BASIC is a rather slow environment for handling real-time problems, such as a mass of data arriving at high speed. The default size of BASIC's serial receive buffer is 256 bytes, so in attempting to receive 10 kbytes of data arriving at 9600 Baud the user's program needs to be able to read the characters from the buffer at a rate of about 960 per second to avoid getting an "INPUT BUFFER OVERFLOW" message. Even on a fast PC/AT this cannot be done. However, the size of the receive buffer may be increased to up to 32k (by loading BASIC with the /C option, eg. BASIC /C:10000 to specify a buffer size of 10k), so unless a vast amount of data is involved it is usually possible to arrange for BASIC to handle fast input. A compiled BASIC such as QuickBASIC runs so much faster than the interpreter that handling serial IO is usually not a problem.

The range of data parameters available through the operating system and through high level languages is as follows:

    Baud rate: 110, 150, 300, 600, 1200, 2400, 4800 and 9600
    Parity detection: N(none), O(odd), E(even)
    Data bits: 7 or 8
    Stop bits: 1 or 2

The serial port provides the most readily available and lowest cost interface for the PC, and its activities are at least partially handled by software routines available within the operating system or ROM BIOS. Many applications packages are provided with software drivers for handling very useful peripherals - such as a Postscript printer. In short, the serial port provides an almost ideal interface point. Certainly all the laboratory interfaces discussed in chapter 7 could be adapted to operate with serial data rather than the parallel user port system described. What then is the problem? In my experience serial data transfers are too slow for all but the simplest applications. Running a serial laser printer to print text is fine. Using it for complex graphics or scanned images is quite another matter; the actual volume of data is so large that it can take many minutes for the data transfer to be completed. It is not surprising that most scanners are equipped with bidirectional Centronics interfaces rather than RS232-type serial ones. For many laboratory applications involving multifunction interfaces, the required rate of data transfer is larger than a serial interface can handle. Operating a dynamic analog input interface may require several hundred byte transfers for a single reading. Even at 9600 Baud a serial interface cannot compete with the much simpler parallel user port approach. Of course, if the PC serial ports could operate at 100 kBaud that would be a different matter.

9.7 A microcomputer/RS232 adaptor

While most modern microcomputers are equipped with at least one serial port, relatively few are fitted with more than two. Furthermore the serial ports on a PC are interrupt driven and the operating system only has two interrupt lines available for handling these ports, so that adding, say, a dozen serial ports can present a problem. In spite of its limitations the simple inter-device wiring which may be employed for a minimal serial connection can make multiple serial ports attractive in some circumstances; for example, for collecting data from many widely separated monitoring devices in a plant. For this reason we will examine one method of implementing a minimal serial port for a computer which provides access to an 8 bit data bus or a parallel user port of the kind discussed in chapters 6 and 7. We have used the approach described with Commodore (PET and 64) and BBC micros, and with PCs fitted with the user-port adaptor described in chapter 6. The method is illustrated along with its extension to four serial ports, although the approach could be readily extended to much larger numbers.

A simple two-way data communication adaptor for converting 8 bit parallel data into RS232C compatible serial signals is shown in fig 9.8. The brain of the circuit is an LSI IC known as a Universal Asynchronous Receiver and Transmitter, a UART, in this case the 6402, although a host of similar circuits is available. UARTs are particularly useful because they require only a clock signal (which runs at 16 times the required Baud rate of the serial input or output), 8 bit input and output connections to parallel data busses, one serial input and one serial output and a few control signals, to provide both-way parallel/serial conversion facilities. Furthermore the format of the serial data, the number of data bits, the number of stop bits and the parity may all be determined by setting logic levels on the control connections, and the UART provides a number of status lines which can be monitored to check when data has been transmitted or received, or whether any error has been detected. The various latches used in the circuit may be addressed using any of the address decoding techniques discussed in chapter 5 or 7, and the decoding circuitry is not shown in fig 9.8. In our circuit the clock signal is supplied to both the received data clock input and the transmitted data clock input (pins 17 and 40 of the UART), but this is not a limitation as the

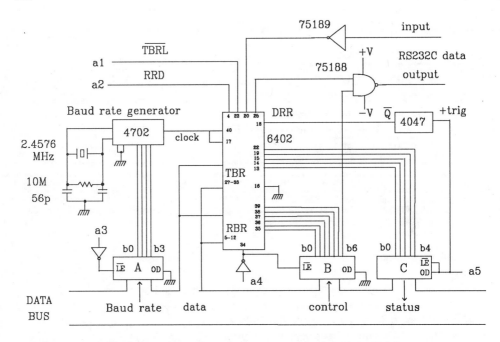

Fig 9.8 *A micro/RS232C adaptor based on a 6402 UART.*

clock rate can be changed easily by latching into latch A (address a3) a code which sets the Baud Rate Generator IC, the 4702B, to provide a clock rate equivalent to any of the standard Baud rates. Byte values of 2-15 (0 and 1 are used for other purposes, see 4702B data sheets) allow Baud rates of 50, 75, 134.5, 200, 600, 2400, 9600, 4800, 1800, 1200, 2400, 300, 150 and 110 respectively.

Similarly codes latched into latch B (address a4) determine the format of the data according to the following bit assignments:

b0   parity inhibit (PI); when b0=1 parity generation or checking is inhibited. This overrides even parity enable (b4).

b1   stop bit select (SBS); determines the number of stop bits for a given number of data bits according to the following rules:

b1 value	data bits	stop bits
0	5	1
0	6-8	1
1	5	1.5
1	6-8	2

b2 and 3 character length, ie. number of data bits, according to:

b2 value	0 1 0 1
b3 value	0 0 1 1
no. of data bits	5 6 7 8

b4 even parity enable (EPE); a 1 selects even parity, a 0 selects odd parity. Even parity requires an even number of 1s in the data+parity bit sequence. This code is only operational when b0=0.

b5 master reset (MR); a 1 clears all status bits. Used to reset the status bits after an error. Note that this bit must return to a 0 before normal operation is resumed.

b6 0 selects "break" output (a sustained 0 level) on the serial data transmission line. Should be normally 1. Note: not connected to UART.

Latch C (address a5) allows a number of status bits to be read by the computer. The bits have the following meanings:

b0 PE, a 1 indicates a parity error, ie. a received character did not conform to the parity specified by b4 of latch B.

b1 FE, a 1 indicates a framing error, ie. the received level was not a 1 when a stop bit was expected.

b2 OE, a 1 indicates an overrun error, ie. data has been received before a previous character has been transferred to the receiver buffer register and DR (b3) reset to 0.

b3 DR, data received flag. A 1 indicates that a data character has been correctly received and is ready in the receiver buffer register to be placed on the parallel data bus. This bit must be reset to 0 (by a lo on DRR) before a new character can be received without an overrun error.

b4 TBRE, transmitter buffer register empty flag. A 1 indicates that the transmitter buffer register is empty and ready to receive new contents.

Note that addressing latch C also triggers the 4047B monostable, which pulses the DRR (data received reset) line of the UART. This clears the data received flag (DR) so that a new character can be received in the data receiver register. Latch C must be addressed before each character is "read" by addressing the adaptor with address a2. In operation for receiving data latch C is examined repeatedly until a 1 at b3 indicates that data is ready for reading from the receiver buffer register. Address a2 enables the output of the receiver buffer register and causes the buffer contents to be loaded onto the parallel data bus.

Data for serial transmission is latched into a transmitter buffer register from the data bus by addressing the adaptor with address a1. When this address line returns hi transmission of the character begins. Note that the UART uses conventional 0 and 5 V logic levels for all signals. Transmitted data (leaving the UART from pin 25) passes through an RS232C driver

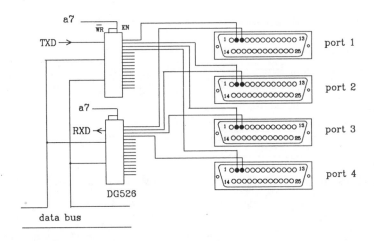

Fig 9.9 *An RS232C connector selector for the micro/RS232C adaptor.*

(the 75188 NAND-type gate) where it is converted into +/-V levels (with -V = 1 logic). This IC therefore uses supply levels which are different from the others in the circuit and which can range from +/-5 V to +/-15 V. Received data is converted from RS232C levels to TTL logic levels by passing through the 75189 (inverting) receiver before reaching the UART. Examination of the data sheet for the 6402 UART will show that the above circuit could be implemented without the need to provide latches B and C (as pin 16 provides and output disable point for the status lines, and pin 34 accepts a signal to latch the levels of the control lines). However, it is convenient to use the octal latches as shown because their spare lines (b7 on latch B and b5-7 on latch C) may be used for the transfer of levels from control pins of the RS232C connector. (b6 on latch B may also be used for this purpose if the "break" facility is not required.) If these spare lines are used for this purpose it is, of course, essential to provide level-changing ICs in each one (not shown in fig 9.9) to convert from the RS232C control levels to 0/5 V levels. Typically the DTR and DSR signals may be incorporated in this manner.

One of the circuits from fig 9.8 (or indeed any serial port) may be used to drive up to 16 RS232C connectors using the selector circuit shown in fig 9.9. A pair of 16-channel analog multiplexors are used to switch the transmitted data (TXD) and received data (RXD) signals to the desired connector (only four are shown in fig 9.10). The connector is selected using a 4 bit code from the data bus and both multiplexors are activated by the same address line (a6). Additional multiplexors could be incorporated to select control line signals (such as DSR) if required.

9.8 Communication using modems

Looking back to fig 9.2 will indicate that the RS232C system was designed to allow terminal equipment to interface to modem equipment. This interface is required so that when the modems are connected together the terminals can communicate with one another. Modems permit the several signals provided by and required by terminals to be compacted to

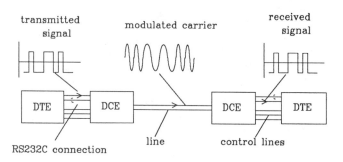

Fig 9.10 *A modem interconverts between the RS232C-level signals and the*
modulated carrier signal required for remote communication.

a form in which they may all be passed over a single pair of wires formed by the inter-
modem connection, the "line" - generally the telephone line. The form of signal passed
between modems is of course nothing to do with the RS232C standard or its newer variants,
but is specified by standards laid down by the owners of the line, many of whom adopt one
of the CCITT standards. The precise details of these standards are different in different parts
of the world, and in any case would be of interest only to the manufacturers of modems. For
this reason our discussion of communications will be restricted to the general principles
rather than the detail which may apply in any particular country.

The only signals which may be sent along telephone lines by users are ac waveforms of a
controlled range of frequencies. Thus a modem interconverts between the signals and levels
required by its RS232C connection to a terminal and a number of ac waveforms allowed on
the telephone lines. Essentially a modem modulates a transmitted carrier waveform of one
frequency by adding one or more other frequencies which correspond to the 0 and 1 levels of
the RS232C signals representing the transmitted data. It also needs to demodulate a received
signal to extract from it the 0 and 1 levels of received data. Fortunately it does not need to
send the control signal levels across the line, as these are either produced within the modem
when it is receiving a carrier, transmitting a carrier, or receiving a ringing signal, or (in the
case of RTS and DTR) they are interpreted by the modem as signals to initiate or permit
some action (such as transmit a carrier). Thus the control signals may be regarded as locally
generated and the data signals are those which are modulated or demodulated and sent across
the line as illustrated in fig 9.10.

Modern modems generally implement one additional signal not referred to in our earlier
discussion. This is the Ring Indicator signal (RI) generated by the modem when it detects an
incoming ringing tone on the telephone line. Many DTEs ignore this signal line, including the
MS-DOS on the PC. But user's software and many packages designed for handling modem
communications (comms packages such as Datatalk, Crosstalk and Smartcom) may monitor
this line so that they can respond to incoming calls. A typical connection between a PC and a
modem modem is shown in fig 9.11, where it may be noted that all 9 lines available on the
PC\AT serial port are used. The control lines function in the manner described earlier: DTR
is set (to true) whenever the terminal is operational and a modem will terminate

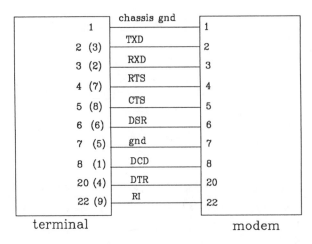

terminal   modem

Fig 9.11 *The complete connection between a PC and modem for telephone communications, showing the ring indicated (RI) line. The numbers in brackets refer to the pin numbers of the PC/AT 9-pin connector.*

communication if DTR goes false; DSR is true when the modem is operational; RTS is set by the terminal when it wishes to transmit data, although data transmission is only permitted when the modem has set CTS; DCD is set by the modem when it is receiving a carrier signal. In half duplex mode the modem makes DCD false whenever it detects RTS true.

In operation the modem transmits (or receives) a modulated carrier waveform over the line. The type of modulation and the frequencies used are specified in the communication standard to which the modem must conform. For example, the CCITT V21 and V23 and the BELL 103 and 202 standards specify modulation using the FSK technique (Frequency Shift Keying), which requires that the range of frequencies transmitted is approximately 1.5 times the Baud rate. In the UK a frequency range of 3100 Hz (from 300 to 3400 Hz) may be used on the telephone network, while in the US a range of 2300 Hz is permitted. For a modem to transmit at 300 Baud using V21 or BELL 103 it must use a frequency range (bandwidth) of 450 Hz. Similarly to receive at 300 Baud also requires a bandwidth of 450 Hz. To transmit and receive at the same time (ie full duplex operation) requires that both 450 Hz channels are in use simultaneously, so they must occupy different frequency regions and in fact should have a small gap (channel separation) between them to prevent interference. So a full duplex 300 Baud system requires a bandwidth of around 1000 Hz and presents no problem to the telephone system.

Unfortunately repeating the calculations for a 1200 Baud link indicates that around 3700 Hz bandwidth would be required, and this is outside the range of frequencies permitted in either country. One approach to overcoming this problem works moderately well for systems in which one end of the link sends small amounts of data while the other sends much larger amounts - typically the case when a terminal user accesses a large public database; the user sends a few characters at his normal typing rate, while the computer holding the database responds with screenfuls of data. In this case it may be acceptable to have one Baud rate for data passing from the user to the database and a different Baud rate for data flow in the other direction. In the UK a split rate of 1200 and 75 Baud has been widely adopted and may be fitted into a bandwidth of around 2000 Hz.

An alternative approach is to use a different communication standard. Much 1200 Baud full duplex communication is now carried out using the CCITT V22 and BELL 212A standards (which are similar but not identical) using Differential Phase Shift modulation (sometimes referred to as four-level phase-shift keying). This technique encodes data bits at two per Baud, so that 1200 bits per second are actually transmitted at 600 Baud. Furthermore an enhancement to V22 is V22bis which uses Quadrature Amplitude Modulation and allows four data bits per Baud to be transmitted, so that 2400 bits per second may be sent at 600 Baud.

Using V22 or V22bis is certainly attractive as it permits data to be passed over the telephone lines bidirectionally at tolerable speeds. However, the rather careful extraction of digits from closely related frequencies requires much more expensive circuits inside the modem than was the case for FSK modulation, so the modems are considerably more expensive. Even so the higher speed modems are more prone to errors through noise on the telephone lines in the UK, and on a bad day communication at 300 Baud can prove faster than that at 2400.

Listening in with a conventional telephone will demonstrate the range of beeps, cheeps and burps characteristic of modem communication. Modems are available for a wide range of communication standards, some for private lines rather than the public telephone system. Most modems are equipped with the 25-pin D socket of the RS232C standard, although for PCs (particularly the lap-top portables) modem cards which fit inside the PC (and so do not actually need an RS232C connector) are increasingly common. While some modems operate to a fixed standard (eg the V21 or BELL 103 300 Baud full duplex) or have no "intelligence" of their own, others may allow several communications standards to be used and offer many useful functions to be controlled via character strings sent from their local terminal (such as dial a number, hang up, generate a local echo, or select touch-tone or pulsed dialing). Some modems may be able to originate calls to other modems (originate mode - remember the two bands of frequencies are required for full duplex operation, so the originating modem will transmit on one band while the answering modem transmits on the other band), and permit other modems to call them (answer mode). Others may only be able to originate a call or (in the case of some mainframe multiline modems) answer a call. Some modems may be able to dial a telephone number (auto-dial) and answer a ringing signal (auto-answer - typically by sending a carrier to the originating modem and waiting for another carrier to be returned) according to some prescribed standard such as the CCITT V25, and some are "smart" and may adjust their operation to one of several possible carrier standards (and therefore Baud rates) received from the line.

One of the first good quality modem modems to come into widespread use was manufactured by Hayes Microcomputer Products Inc., and the Hayes Smartmodem has a command structure which has become a de facto standard for most other good quality modems, although (in Europe) often enhanced to support split Baud rate operation. Modems

Table 9.3   Examples of commands which may be sent to Hayes-compatible modems

Command	Function
Dxzy	the modem is set to originate mode and then dials the number xyz
,	forces a pause of one second (generally used in a dial string to allow a delay for a secondary dial tone)
T	use tone dialing
P	use pulse (loop disconnect) dialing
A/	repeat the last command (generally for redialing)
;	return to Local command state after dialing (used after Dxyz)
Sn	set modem to answer mode after nth ring
A	seize the line
Cn	turn carrier on (n=1) or off (n=0)
En	local echo on (n=1) or off (n=0)
Hn	seize (n=1) or drop (n=0) line
In	request modem (n=0) or modem ROM (n=1) identity code
Mn	turn internal speaker on (n=2) or on until carrier detected (n=1)
O	return online
Z	full modem software reset (restores all default values).

which operate with "Hayes compatibility" may in fact function in two states: the On-line and Local command states. In the On-line state data sent from the terminal is normally transmitted over the line, but in the Local command state specific commands may be sent from the terminal to instruct the modem to perform certain functions - such as dial a number, generate a local echo, etc. These commands are sent to the modem as a string of normal ASCII characters, and so may be typed by the user on his terminal or may be sent by software running on the user's terminal (eg the user types dial and his program sends the dial command to the modem). A list of some of the commands commonly available is given in table 9.3, although it should be noted that all commands must be preceded by the capital letters AT which may be followed by a space.

9.9 File transfer protocols

Controllable modems can be very useful. But before getting carried away with the possibilities, it is worth remembering that most of their actions are taken in response to commands from the attached terminal. Some such commands will be given manually by a user, but most are (and are best) provided by software designed to handle the required communications. Generally it is the responsibility of the user's software to ensure that data communications take place without error, and this requires not only that the hardware link between the two terminals is standardised, but also that the software operating in the two terminals adopts a standardised method for detecting and correcting transmission errors.

By far the most widely used approach to error detection and correction is a file transmission protocol called XMODEM (developed by Ward Christensen in the late seventies). Under the XMODEM protocol one computer sends data to another in the form of "blocks" of many characters, and each block contains a start-of-header character (ASCII 01, SOH), a one-byte block number (which cycles back to 0 after 255), a one-byte ones complement of the block number, 128 data bytes, and ending with a single-byte checksum calculated from the block's contents. When received at the other computer (which must also be using the XMODEM protocol) this computer can make the same checksum calculation and compare its results with the checksum received along with the block. If the checksums differ then the protocol allows the receiving computer to request that the same block is re-sent, until either the checksums do agree or the controlling software aborts the transfer. Of course, if the error rate is high (as it can be when using a high Baud rate on a noisy telephone line) then the rate of successful data transfer can be rather slow as blocks are re-sent time and time again. XMODEM is not a particularly efficient transmission protocol, but it does have the advantage of being provided in most comms packages. For this reason there is a reasonably good chance that any two comms packages running on different computers will be able to swap files.

The XMODEM protocol requires the file-receiving computer to transmit a negative acknowledge character (ASCII 15H, NAK) to indicate that it is ready for the first block of data, and an acknowledge character (ASCII 06, ACK) when the block has been successfully received. In the event of a checksum error the receiver transmits a NAK character to indicate that the last block should be retransmitted. After the last block has been sent the transmitter sends an end of transmission character (ASCII 04, EOT).

One of the problems with XMODEM is that the single byte allowed for the checksum is not sufficient to trap all transmission errors. An enhancement to XMODEM, which overcomes this difficulty is the CRC (cyclic redundancy check) option. The CRC enhancement allows for a two-byte CRC value (generated using a polynomial formula) to be sent instead of the single-byte checksum, and is claimed to detect 99.9% of errors.

There are many other transmission protocols, including Kermit, Ymodem and Compuserve-B, and many are much more efficient than XMODEM. Some allow the name of a file to be sent in advance of its contents, so that the received data may be stored in a file with the same name. Some allow the receiving end to request a named file, and many do manage a higher rate of successful data transmission than XMODEM. The problem is that many of these protocols are different from one another, so that it becomes necessary to ensure that both computers are running the same comms package to ensure success. Datatalk is a comms package which I have used regularly for a number of years, and it offers both XMODEM and Datatalk protocols. Unfortunately relatively few of the people I exchange files with have Datatalk, so most exchange has to be done using XMODEM. Kermit has come into widespread use in the UK, partly because it is freely available at most computer centres, although until recently it was not supported by most commercial comms packages. With multitasking now a reality for the PC/AT, it is a relatively simple matter for me to leave a computer running an instrument for weeks on end while running a comms package as another task. I can then call that computer from the comfort of my home, copy over the filed results and send back files of new instructions.

CHAPTER 10

SYSTEM DESIGN AND CASE STUDY

In designing any laboratory system (or indeed anything else) it is sensible to follow a logical procedure. The importance of design becomes apparent after one's first creation, but the need to devote time and energy to design and the necessity of following a design procedure continues to impress me with increasing force after many creations. In lecturing on the topic to both students and practising laboratory scientists I continue to find that the standard reaction is: "Cut the philosophy - let's get on and make something." The number of times that this is followed, eventually, by "We can't make it do that - we would have to rewrite the whole thing." or "We can't measure that as well - there are no spare address lines," convinces me that this chapter should be the most valuable of all.

In the present context design is important for an additional reason. Most instrumentation based on microcomputers will have both hardware and software components. While some individuals will undoubtedly have the ability to design both aspects of the overall system, in most cases the final result will derive from the efforts of a team. Properly thought out and discussed design will ensure that not only is the impossible not attempted, but that the members of the team will all be trying to construct the same instrument, rather than several different images from their own imaginations.

In this chapter we will examine one approach to designing instrumentation. Naturally it is not the only approach, but it is one that I have found reasonably effective. To illustrate the operation of the approach an abbreviated case study is included, and this draws extensively on the principles, devices and circuits discussed in earlier chapters. More detailed accounts of some of the hardware and software aspects of the instrument involved may be found in the issues of Laboratory Microcomputer, and results obtained with the instrument are discussed in the Journal of Automatic Chemistry.

10.1 An approach to system design

A laboratory instrument is best designed by working through a number of individual design steps. The steps follow a logical sequence, although different designers would probably itemise the steps in different ways (indeed, in an attempt to follow more closely the way in which I work I have modified the order of the steps since the first edition). One approach suitable for systems which involve a substantial computer component utilises the following steps:

1. Determine system objectives.
2. Assign priorities to system performance.
3. Define system specifications.
4. Create operational concept.
5. Determine system components.
6. Devise software overview.
7. Outline design/specify hardware components.
8. Decide whether to proceed.
9. Design main software flowchart.
10. Design in-house hardware.
11. Draw software flowcharts.
12. Write software.
13. Test components and assemble system.

Although it is not possible to cover every eventuality, we will discuss each of these steps in turn before examining case studies of specific instruments designed in our laboratories.

Step 1. Determine system objectives

By the time he comes to design a system the designer probably has a fair idea of what it is supposed to do. Nevertheless it is essential that he disciplines himself to list precisely what the objectives are in designing and producing the system. ALL the objectives must be listed. For example, in designing a system for recording chromatograms, a primary objective is likely to be an instrument which will record the chromatograms. But the cost of the system may be important. It may be that an objective is to design a system which will cost less that £5000. If so, then this must be listed. It will have a significant bearing on decisions taken in subsequent steps. It may be that one of the objectives is to produce chromatograms on paper with the mass of the separated components printed alongside each peak, or from both gradient elution (ie varying eluent) or isochromatic elution (constant eluent) runs. It may be that different types of detector may be incorporated into the system in the future. If so this must not be left off the list, as it may profoundly affect decisions on interfacing hardware. It may be much harder to incorporate changes to the system later, while it should be relatively easy to build in a certain amount of flexibility and expandability at this stage if one has some idea of the directions which future modifications are likely to take. Is it likely, for example, that the users may wish to compare results on screen with results obtained six months earlier, or with results simulated by a program they have operating on an existing computer?

Step 2. Assign priorities to system performance

Once the system objectives have been determined it is necessary to assign priorities to the several characteristics required of the system in achieving those objectives. For example, is ease of use by relatively untrained personnel more important than speed or accuracy? Or is security of recorded data overwhelmingly important? In most cases the requirements of these various aspects will conflict with one another. The priorities assigned at this stage will be referred to in future steps by all associated with the project so that sensible decisions can be taken when a conflict needs to be resolved. So it is essential that priorities are assigned - perhaps by using a numbering system of 1-10 - and that all involved should be informed of the priorities at the earliest opportunity. Any safety aspects should be extracted from those who understand the nature of the system and assigned an appropriate priority.

Unfortunately the system designer may not be in a position to assign priorities, particularly if is he designing an instrument for others to use. Furthermore, other people will probably have different priorities for different aspects of the system. Thus the potential users may regard ease of use as number 1, while the laboratory manager may regard accuracy, and the company security as the overriding factor. However, somebody must be responsible for specifying priorities right from the start (and in my experience it is generally he who pays the bill).

Step 3. Define system specifications

At this stage the designer should know what the system must do and have determined the priorities for the system's characteristics. It now becomes possible to list the design specifications for the system, ie. what inputs the system will need to accept, what output(s) it will need to provide and in what form; what parameters need to be varied and over what range, etc. This list should include all specifications, including those which may be required only as subsequent additions to the system. For example, it may be that a system will be used initially with a chart recorder output, but with the hope that one day a laser printer might be used for hard copy. Clearly if such information were not available while the specifications were being determined, substantial difficulties could arise at a later stage.

Generally it is advisable to determine both desirable and minimum specifications, so that there is some room for compromise in subsequent steps. Thus it may be desirable that a spectrometer covers the wavelength range 190-800 nm, while a minimum acceptable coverage may be 200-500 nm. The interplay between these specifications and cost, complexity, size, timing (spectrometers often pause to move optical components during a scan), may be of vital importance when the decision about whether to proceed is taken. It would be a pity to abandon a project simply because one aspect had been overspecified on "wouldn't it be nice" grounds. The priorities assigned in step 2 will undoubtedly influence the difference between the desirable and acceptable specification figures. Any safety aspects relating to the system must have suitable specifications defined (for example, time to turn off in an emergency, or the level of leakage of microwave energy from a cavity).

In subsequent steps one would expect to start with a design specification equal to the desirable specification, altering the design specification if necessary, but never going below the minimum acceptable specification. Many factors could force compromises in specification at several subsequent points, even though such problems may not be apparent at this stage. For simple systems in which all stages are being undertaken by one individual, modifications to the design specification should be easily adopted. But for more complex systems in which several people are involved in different aspects of the design it is vital that all should be aware of the design and minimum acceptable specifications, and that any modification to the design specification is quickly notified to all.

Step 4. Create an operational concept

Different people approach the activity of system design in different ways. At this point some would determine what components would be needed to achieve the specifications defined in the previous step, and later blend these components together to form an instrument. My preference is to attempt to envisage a concept of how the end product will appear to its users, and then to decide on the components which will achieve that vision. For the purposes of the present discussion we will assume that a computer based instrument is envisaged, although it is important to bear in mind that some objectives can still be achieved without the participation of a computer. Building an instrument around a computer when the job could be accomplished with a simple electronic circuit is an excellent way of passing large amounts of time, but it is not a particularly efficient use of resources.

At this stage the designer knows what inputs will be available and what outputs will be required (although in most cases he will not have determined precisely what signals will be required to implement measurement and control functions - this will come later). He should now reflect on what interactions between the user and the instrument will be most desirable for the end product. Should the system be totally controlled from the computer keyboard? Or should other parts of the system have knobs, buttons and switches? Does the system have more than one function to perform? In what way should the user select between these functions? Should all aspects of the system be apparent to the user, or should some (eg. calibration and testing) be hidden from the user and only available to the "engineer." If data analysis is be carried out on data collected by the instrument, how and when will this be done? While the instrument is collecting other data? Or will the instrument be unable to function while its computer is used for such analysis? How will hard copy output be obtained?

Some aspects of these considerations will be influenced by the priorities for system characteristics laid down in step 2. However, the creation of the operational concept remains the most difficult part of the design process. It is important not to start on this step before the previous steps have been completed - otherwise one can end up with a superb instrument, but one which does not do what is required. Also it is difficult to avoid a preconceived operational concept influencing one's attitude to objectives and priorities, and in most cases this will be highly undesirable. A designer should take particular care to keep an open mind when devising an operational concept. It is all too easy to work for a while on one concept, and then to become very defensive about it, trampling over all objections and becoming more and more determined to maintain it as one's own creation. I usually make a point of devising two or three quite different operational concepts for any instrument in an attempt to overcome any natural desire to concentrate on the first one that comes to mind. I then try to determine which concept most precisely and simply meets the criteria laid down in the previous steps.

The concept created at this stage must be modelled, ie. set down on paper or made up in a dummy instrument. This is particularly important where other people are to be involved in the implementation - again because it is essential that everbody concerned is working on the same instrument. A clearly defined concept avoids components added in subsequent stages appearing incongruous, such as the odd switch appearing on an otherwise keyboard-controlled instrument. It will also help in the outline of software - thus it is unfortunate if a system is largely menu-controlled, but requires a complex key sequences to operate a specific function of the instrument.

In some systems safety aspects will require careful consideration, although they would have been identified when the priorities were laid down. Whether these can be dealt with adequately by software or hardware must be determined, and in serious cases consultation with a safety expert may be required. Safety aspects cannot be dismissed lightly, as it is all too easy to devise automated instruments which are incapable of detecting a hazardous situation which would be obvious to a human operator, and all too easy to have the system unstoppable until it is too late. [For example, one of the instruments built in our laboratories contained a flame ionisation detector driven by a flow of hydrogen and air. The system also contained facilities for automatically igniting the flame. As the instrument was housed in a 1- metre cubic cabinet, attempting to ignite the flame when the cabinet was filled with hydrogen and air could lead to a spectacular explosion. It was necessary to design in adequate safety controls, to (among other things) turn off the hydrogen flow if the flame was accidentally extinguished.]

Step 5. Determine system components

Once the operational concept has been decided it becomes possible to determine what components are required to turn the design specification into a practical design which fits the concept. This is a time consuming aspect of design, requiring a good deal of work and a number of answers (from both the designer and others). Which transducers respond to the properties of interest? What kind of signal converters are required to generate computer-useable outputs from the signals provided by the transducers? Exactly what type of information presentation devices may be used to produced the required outputs? Meters, digital displays, video screens, a chart recorder, graphics plotter or printer? What kind of computer would be needed to handle the volume of data and the resolution of the graphics display? What about disk storage, floppy or hard disk? Would non-volatile RAM be more suitable for the environment in which the instrument is to be used?

At this stage the designer needs to acquire detailed specifications of any commercially available components he wishes to use, including their cost, availability and compatibility with other components. Ancillary items which seem to be essential for the required operation of commercial components must also be noted. The availability of recently announced items may be worth double checking, as this can turn out to be rather different in reality from the promises of the advertisements.

Once specific major components have been determined (such as chromatographic detectors and computers) it is necessary to examine what kind of interfaces will be required so that the specified signals of one device may be transferred to another. Are any of the major components available with built-in standard interfaces (such as GPIB)? Would a commercial general purpose, multifunction interface be suitable for the job? If so, how much "intelligence" should it have, if any? What components will need to be produced in-house, and how much will this cost? Even worse - how long will it take?

Software is always a major component of computer-based instrumentation, and is just as real and more frustrating than the hardware. Who is to provide the software? Will a commercial software package (such as Lotus Measure) be suitable, or will someone need to write the software required? The designer may know BASIC and even like the idea of writing his own software. But is this realistic? Software will always need support once it is in use. Will the designer be able to support it in a year's time? Or will it (and the whole instrument) become a white elephant after the designer leaves and nobody can unravel the workings of the program which won't run? Would it be better to talk the user or someone else into writing his own software, perhaps with the aid of a few routines which handle the interface activities? If speed or security were high priorities, then perhaps a compiled language would be more appropriate for the software. How much is a compiler for the computer selected? Who knows enough about programming to spend weeks or months writing a substantial system and providing the documentation necessary for somebody else to be able to modify or improve the system in the future? Is there any mathematical apparatus which would be required in achieving the system's objectives (eg. Fourier transforms), and if so are these fully understood - including the implications of programming routines for the required arithmetic precision. To put it bluntly, is the design specification likely to be jeopardised by the wrong choice of programmer?

At the end of this step the designer should be able to produce a block diagram of the planned system, showing the major components, including software. For complex systems it is sensible to include within each block details of a) who is supplying the component, b) the expected delivery time, and c) the estimated cost. In all probability the most difficult part of this is in estimating the cost and delivery time of the software component, and, in practice, reasonable estimates are unlikely to be available until later in the design process (and good estimates, probably not until after the project has been completed!). However, if the software writing is costing real money then it is important to realise that it will probably be one of the major costs of the project, because large amounts of time will unavoidably be devoted to planning, writing, testing and correcting the software whoever writes it.

Step 6. Devise software overview

If in-house software is to form a major component of a system then it is as well to outline the software's roles and components before a decision is taken on whether the project should proceed (ie. before deciding that the hardware should be purchased or made and the detailed software written). In part this provides some insurance against designing systems which cannot possibly be implemented. In any event, it allows the designer to expose his operational concept to the realities of software limitations and, hopefully, to someone who is experienced in designing software. An overview flowchart may be constructed to show the number of programs required to implement the functions to be provided by the instrument. Definitions of the actual requirements of each program can be set down, the mechanism by which they may communicate with one another may be specified, and the nature of any important data structures can be determined. The language or languages in which the software is to be written should be decided, generally on the basis of the expertise available, the facilities provided, ease of use of mathematical or graphical subroutine libraries or a floating point coprocessor, etc. In many cases there is no reason to avoid deciding that different components of a system should be written in different languages. For example, it may be sensible to write a data collection system in C, while a pattern matching search facility might be better provided via Prolog. Of course it is essential to ensure that suitable compilers are available, and that programs in one language can pass parameters to routines in the other. By considering all these aspects it becomes possible to ensure that the software could actually be implemented on the chosen system, given its memory size, disk facilities and languages available.

A software overview chart will specify for each program what it actually does in terms of inputs and outputs and data handling, and what form data is to be stored in (bytes read from an interface or floating point variables). Any other decisions about what needs to be stored or used should also be noted. For example, if a system is to record a spectrum, do we need to store both the intensity value and the wavelength for each recorded point, or will the intensity be sufficient if taken with a wavelength starting point and step size? Is temperature important, and how is its value obtained by the software? If mathematical adjustments (corrections) are made to data, do we need to store both corrected and uncorrected forms? If the correction factors are transducer dependent (such as the conversion efficiency of a photomultiplier tube as a function of wavelength), which program will allow these factors to be changed if the transducer has to be replaced? Should the data files be stored on the same disk as the program files, or will this slow everything down too much? Are users to have their own floppy disks for their own data, or can everything go onto a hard disk with the user given an option to copy to floppy? Is it likely that the software will be required to operate within different environments (eg. with monochrome and colour graphics display adaptors)?

At the end of this step the designer should have a reasonably clear idea of how many programs are required, what language(s) will be used, what the principal data files will hold, and roughly how much memory any particular program is likely to need. He will have a chart on which he shows which programs will access which data files, which programs will access which peripherals and which will access interface units. He will also have notes on what each program needs to be able to do, and he will be confident that what is required is possible to achieve. With luck he will be in a position to obtain a more accurate estimate of the cost which will be incurred in producing the software.

Step 7. Outline design/specify hardware components

This step involves performing for in-house hardware components much the same kind of planning that the last step performed for software. And for much the same reasons: to ensure that what has been specified is practical, and to outline in block diagram or sketch form the way in which these items of hardware may be fabricated. In most cases this will involve items such as equipment housings or power supplies, where there is little doubt about the feasibility, but the capacity may need checking, or interface units, where questions such as the type of interface connection (serial or parallel), the number of addresses, the speed of operation and the characteristics of the conversion may need to be examined.

Of course it may be that a standard interface system (such as the GPIB) is already available on some major components of the system, and this would be likely to influence decisions made at this time. However, if this is not the case then now is a good time to determine what circuits will be used for such items and how they will need to communicate with the computer. Is an addressable interface required, or are there enough IO ports available (and not reserved for other purposes) on the micro to handle the individual devices? Do all the devices need to operate at high speed using synchronous techniques, or could some be more economically implemented using a user port system? What is the best way of powering the interfaces? Using the computer's power lines or constructing a separate power supply? What is the most economical form of construction; stripboard, wire wrapping, IDC techniques or printed circuit cards? Should the circuits be on one large card, or distributed among several smaller ones which are interconnected? What are the key points in the circuits for protection against interference, and what is the worst-case disaster which could occur to computer or instrument if one of the components failed? What connectors and cabling will be required, and over what distances would the system components need to function?

At the end of this step the designer should have rough sketches of any hardware components he will need to have prepared in-house. These include rough circuit diagrams (ie not down to resistor values and pin numbers) for any electronics involved, particularly the computer interfaces, to ensure that the circuits are compatible with one another. This level of detail may at first sight seem premature, but it is vital that the designer can be sure that these small devices can be constructed to work together and with the computer. Apparently minor errors can prove very costly: some TTL circuits may not be available in the 74LS series, and yet a CMOS output cannot drive a battery of standard TTL inputs; some ICs respond to level changes rather than static levels; some circuits produce or use hi busy signals, while others use lo levels. Having to change a type of IC once a printed circuit board has been designed can be expensive.

At this stage it will be possible to provide a reasonably accurate estimate of the cost involved in the production of in-house hardware. Frequently the cost of cases and connector will exceed that of the ICs, although both will be small compared to the cost of circuit board design or one-off fabrication. However, in general the cost of design or fabrication can be estimated adequately from circuit diagrams.

Step 8. Decide whether to proceed

At this point the designer is equipped with plans and estimates for all components of his system. He can now state with some confidence whether or not the system will be able to meet the objectives, how the performance priorities have been catered for, and how the system will appear to the user. He has a fair idea of how much the system will cost, when the parts will be available and how long it will take to construct the system. (It is good policy to double any estimate on the software development time.) While the designer has undoubtedly invested a considerable amount of time in the project so far, and some time has probably been invested by the software and hardware experts, no major expenditure has been undertaken. It is still not too late to start all over again!

Now is the time for somebody to take the decision about whether to proceed to commit the money and resources to turning the plan into reality.

Once the decision is taken, the thought that went into the preliminaries should start to pay off. Major components can be ordered, final design and construction of small hardware items can be undertaken, and the software design starts in earnest. Hopefully this is not the time at which the software expert enters the picture; he should have been there since step 3, where he advised on the specification of aspects of the computer system. Subsequently he would have advised on the choice of computer, and on the refinement (or creation) of the operational concept, and he would have had a major influence on the choice of program and data structures in step 6. Undoubtedly he would have been involved in the selection of interfacing components in step 7, at least to the extent of ensuring that the necessary IO signals could be handled in the appropriate combinations and in the required timescales. The software should not be an afterthought in any computer based system; it is one of the major system components. Even though the detailed software design has not started at this stage, the capabilities of software should have already been crucial in deciding what is possible.

Step 9. Design main software flowchart

In general terms it has already been decided what the major functions of the software are to be and how data is to be dealt with. The task now is to divide these functions into manageable chunks, each of which performs a small number of precisely definable functions and can be programmed more or less independently. The first stage of software design should concentrate on how to make this division and what appropriate functions would be. At this stage this may be accomplished by drawing up a main flowchart which shows how the various functions are logically linked together.

This author's preference is for menu driven software, consisting of a master program (sometimes called a steering segment) which displays a menu of options from which one may be selected in a standardised way by the user (typically by using the cursor and enter keys). When a key is pressed control is transferred to a subprogram which may offer menus of its own, but there is always a path for returning to the earlier menu (typically the escape key). In this case the main flowchart will show how the boxes representing the subprograms are linked to the box representing the main program, as illustrated in fig 10.1. In practice such flowcharts should be drawn on large pieces of paper, as a substantial amount of information will be written on to the chart as time goes on. The layout of subprograms will probably bear a resemblance to the overview chart drawn earlier, although there is no reason why a strict correspondence need be maintained.

The main flowchart should have a number of programming details added to it as decisions are made; details which are essential before the design of individual subprograms can be undertaken. If subprograms are to be written in BASIC, then groups of line numbers should be allocated to each subprogram, and blocks of small value line numbers should be set aside for frequently called routines (such as those handling interface reading). If the subprograms are to be written in a compiled language, then naming conventions for labels should be settled. In either case this avoids the difficulty of clashes when all the subprograms are finally assembled together. Any areas of RAM to be set aside for machine code routines should also be noted on the main flowchart.

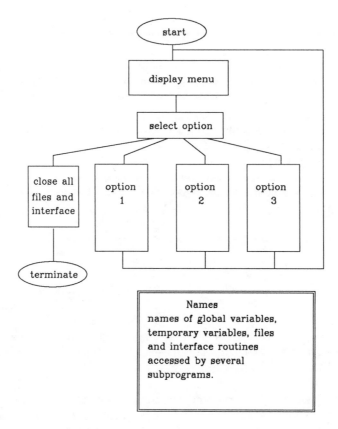

Fig 10.1 *An example of a flowchart showing how subprograms and routines are logically linked together. A main flowchart should also contain information on names which are to be used in several subprograms.*

Each box on the flowchart should contain a few lines of description of the function of the subprogram - lines which could subsequently be included as REMarks or comments within the source code. Also written in should be the names of any disk files which the subprogram will need to access, and, for subprograms, the letters of the alphabet with which variables which are local to that subprogram may begin (this is not a bad policy even for languages in which all variables default to local). In an isolated box (labelled "names" in fig 10.1) should be written the names of global variables which will need to be accessed by several different subprograms, together with a brief note to indicate the quantities represented. Even when subprograms operate largely independently and use parameters to communicate with one another, program listing will be much easier to understand if major variables are consistently named throughout. (It is useful if this list can be included at the end of the main program source code.) Names specifically allowed for short term use (eg. ii as a loop index, temp for a temporary result, etc.) should also be noted, and where several programmers may actually write the code, they should be encourage to stick to these conventions in the interests of consistency. When the final programming language is determined it may be that additional information will need to be added to these boxes to ensure consistency in variable types (such as short or long int) and function types.

The main flowchart should be as complete as possible before work starts on the design of specific program or subprogram flowcharts. It must include details of data structures and file structures, and in my experience it is well worthwhile to build in a certain amount of spare capacity into these (typically a few extra records to contain nothing in particular at the begining and end of each file) to allow for the inevitable afterthought. The chart should indicate which subprograms will require access to files, peripherals and interface units, and identify those which will handle error conditions internally and those which will transfer control to a separate error handler routine. If there are safety aspects to be considered, then the way in which these will be handled must be clearly defined.

When designing a software flowchart for a laboratory instrument in which software and hardware will interact, it is as well to spare a thought for the eventual testing of the system. I have always found it helpful to build into programs one or more variables which can be used as switches, so that programs may be tested without having their associated interfaces actually connected. Of course this doesn't allow interface routines to be tested, but it does allow most other aspects of the software to be quite thoroughly tested in the privacy of one's computer before being let loose on the real system. It will undoubtedly help in enabling graphics displays to be tested in relation to other aspects of the software without having to do this testing on the completed system. Typically my interface handling routines will incorporate a test which will either use the real routine, or will, say, generate a random or calculated number and return. In fact I tend to leave such permanently in system software, as this makes it easier to produce updated versions of the software without requiring extensive testing time on the systems hardware. For some compiled languages switches may be incorporated through compiler options or directives, while for BASIC and QuickBASIC it may be convenient to provide one or more simple variables which can be tested. In any event it is important to note this facility on the main flowchart, so that anybody involved in designing routines will be able to take the necessary actions.

Step 10. Design in-house hardware

Both in-house hardware and software design can now be undertaken, and generally these aspects will be undertaken by different people. Complete circuit and wiring diagrams will be required for any interfacing and ancillary circuits, along with documentation on the principle of operation and any key points in component selection (so that servicing in a year's time does not become excessively difficult). The precise manner in which the circuits are to be constructed, and the mechanical and electrical relationships with other circuits must be determined and the appropriate action taken. For mechanical hardware again drawings will be required and any special materials must be obtained before construction can begin.
If adequate care has been taken over earlier steps then this step should be relatively painless, and the designers and constructors should be in no doubt about the objectives and priorities for the system they are producing.

Step 11. Draw software flowcharts

Given the completed main flowchart from step 9, it should now be possible to draw up flowcharts for the individual subprograms required. Whether this is done by one person or by several will of course depend on the magnitude of the task. In any event, the existence of a detailed main flowchart should help considerably in ensuring that the subprograms are capable of functioning together. For each subprogram a separate routine flowchart will be designed, and this will consist of boxes representing individual routines called by the subprogram, each specifying the inputs provided for the routine and the outputs and/or actions required of the routine. Again it is helpful if the boxes contain short descriptions of the routines' functions so that these can be incorporated as REMarks within the routines. Variable names for all non-trivial variables may be assigned at this stage, and, for BASIC programs, groups of line numbers may be reserved for each routine. It is likely that several subprograms will be able to share routines, and likely candidates for this should be identified while the routine specifications are being defined.

For each routine a detailed flowchart may then be devised. Each has assigned specified line numbers (or label names) and local variable names, and in many cases these flowchart boxes may be completed in BASIC or the chosen alternative language. Any assembler language routines should be charted in as much detail as possible, with label and variable names chosen so that completed coding can be checked against the flowchart in the future (which usually is necessary after the designer has left or forgotten all about the program).

Step 12. Write software

The routine flowcharts should provide all the information necessary to allow the programs and routines to be written in the chosen language. In fact it is only at this stage that one appreciates the effort which has been devoted to the construction of the flowcharts - or, alternatively, wishes that more had been.

Step 13. Test components and assemble system

Wherever possible individual components should be thoroughly tested before the complete system is assembled. In some cases (eg the interfacing units) the component can be tested only in conjunction with others. However, the more testing can be carried out on individual items the less chance there is of having difficulty locating any fault on the assembled system. Furthermore, it is nearly always worth devoting time to the testing of individual components, as this inevitably results in a greater familiarity with the components of the system, and can be of considerable assistance in understanding behaviour if their subsequent interconnection leads to problems (through mutual interference, ground loops, etc.). This is particularly true for software and interfaces, as the nature of microcomputer suppliers is such that they are rarely able to provide detailed technical support on any but the commonest items.

In-house software can take a substantial period of time to test (it can take weeks just to get the code right) and is an aspect which is generally under-estimated when quoting a "ready by" date. If switches were incorporated into the software to enable testing to be carried out before the system was fully assembled, then much of the software testing will be possible before system assembly. However, there will always be some aspects of the software which cannot be tested in the absence of a complete system. Indeed there may be some aspects which cannot be tested without both a working system and a real user.

Once the individual components have been tested the complete system can be assembled and tested, preferably in stages. Assuming that the hardware components work, the main areas of interest will be whether the system is able to meet its design specification, whether the software operates correctly under a variety of circumstances, and whether the user interface (ie the appearance to the user) is as clear and useful as was planned. If everything works perfectly then that is as close as human activities can come to the miraculous. Much more likely is that the software will contain a number of bugs (known as undocumented features) or irritations which will require modification.

My preference is to get software changes introduced in large chunks, rather than to dribble out small changes over substantial periods. Of course, major deficiencies (such as the program crashes immediately after starting) have to be fixed rather rapidly. But where possible it is desirable to allow the system to undergo trials by the user for a week or two, and allow a list of complaints and suggestions to be compiled. At the end of that time a number of software changes may be incorporated at one go, in full knowledge of the interactions these are likely to have, and may be noted on the flowcharts for future reference. Of course a similar list of suggestions will build up from this time onwards and it may be desirable to consider another version of the software after a few months have elapsed. However, it is important to avoid a constant stream of software versions, particularly where the changes will largely involve the user interface (ie cosmetics). Many changes in the user interface can seriously undermine the confidence of users who know little of computer systems, and can create confusion in the mind and memory of the software expert making the changes. Besides there is a certain respectability in adopting the attitude of the major software houses ("We know about that problem; it will be fixed in the next release.").

10.2 Case study: A variable wavelength colorimeter

This study is partly artificial, in that the unit was designed as part of a more complex system. However, the design of the full system filled two large notebooks with notes, diagrams and alternatives, and would be too space consuming for discussion here. The colorimetric detector nevertheless forms a useful case study because it was a complete unit in its own right, and in fact was designed in a hurry when the commercial unit purchased for the larger system failed to perform adequately. The colorimeter is considered in this case study because it was designed, built and tested by the same individual (me), and I made careful notes on the steps involved in anticipation of this second edition. As a colorimeter it could probably stand a fair amount of improvement, and some aspects of its operation may be justified only in relation to other components of the larger system. Even so, it may serve to illustrate the design steps described above.

Note: Absorbance in the present context may be regarded as the logarithm (to the base 10) of the ratio of the light intensity detected in the absence of a sample to that detected in the presence of sample.

Step 1. Determine system objectives.

In this case a visible absorbance detector was required for incorporation into a computer control flow analyser, which was accommodated in a specific cabinet and provided a standard sub-unit housing for the detector unit. The system was required to monitor the absorbance at a specified wavelength of a liquid flowing through a small flow cell, to record the height of the single peak in the absorbance signal produced by each sample, and to be able to change its monitoring wavelength from one value to another within ten seconds. It was also necessary that the cost of repeat units should be under £1000. The unit would be required to operate under the control of a computer (which also controlled the rest of the system), and should be capable of working with a wide range of computers (because the company for whom the instrument was being designed kept changing their minds about which computer system they wanted).

Step 2. Assign priorities to system performance

In this case the assignment of priorities was relatively straightforward and was largely determined by the company requiring the unit. Priority number one was that the unit could record absorbance peaks of less than 0.3% (that's about 0.001 in absorbance units) over the wavelength range 400 - 700 nm, and with a signal-to-noise ratio of better than 2:1. This was assigned a priority value of 1. Priority number two was that the flow cell volume should be less than 50 ml (to avoid undesirable distortions in the measured peak heights), and priority number three was that the system should be capable of operating unattended for periods of weeks (which implied that the computer would need to be able to turn the light source on and off to improve its lifetime). These were assigned priority values of 2 and 3 respectively. Ease of use by relatively untrained personnel was regarded as the next priority, while security of recorded data (in the sense of not losing it if the power goes off) was regarded as worth maintaining to avoid waste of measurements, but was not of a high priority. I assigned a priority value of 4 to the ease of use and a value of 5 to the data security. No other priority features were suggested.

Step 3. Define system specifications

The specifications were drawn up in consultation with the customer. The physical size of the unit's housing was defined so that it would fit into the analyser system. I decided to design for a desirable wavelength range 350 - 750 nm, using the 400 - 700 range as the minimum acceptable range, a light detection system capable of reading to 0.02% (0.05% acceptable) of full scale for the intensity level, a desirable flow cell volume of 30 ml (acceptable maximum 50 ml). A relatively large optical bandwidth (2 nm) was considered acceptable, so a desirable specification of 1 nm was chosen.

After a little probing it became clear that the customer may be interested in the future in having facilities to scan a spectrum of material stopped within the flow cell, so I decided that it would be necessary to design the wavelength selection system so that it could move from one end of its range to the other while intensity data was being recorded and within 20 seconds. It also determined the choice of wavelength selection system, as a filter wheel (which would probably have been adequate for fixed wavelength monitoring) would have been unsuitable for recording spectra.

Step 4. Create operational concept

In this case the operational concept was relatively quickly arrived at through constraints imposed by the existence of the larger system. Essentially the user would define an operating wavelength for the monitoring of absorbance peaks, and would receive in return a graphical display of the variation of absorbance as a function of time for each sample presented to the detector. The software would also locate the baseline and the top of each peak, calculate the peak height and report this as a value printed at a specific screen location as each peak appeared. The system would also be able to print out graphical and numerical records of the data from any number of samples, either while recording or at a later time on demand from the user.

The detector unit itself would have no knobs or switches on its front panel, and a single indicator light to confirm when it was under power. The unit would be wholly controlled by the computer's software and would provide output to the user only via the computer's display or printer.

Step 5. Determine system components

There was no choice over the computer attached to the unit, as it was already part of the analyser system. A PC clone with a 20 Mbyte hard disk and EGA adaptor and monitor was thus a defined component of the system. Two expansion bus slots were available for interfacing, although the existing serial and parallel ports were already in use, the latter being connected to a graphics printer.

A little research among the catalogs soon led to a suitable monochromator (and grating) which could have its pass wavelength set by a stepper motor. Unfortunately there was no way in which the monochromator and stepper motor combination could be fitted into a housing of the specified size, but an examination of the catalog of a stepper motor manufacturer indicated that a much smaller stepper motor could be accommodated. The monochromator suppliers were happy to supply a monochromator and the mounting brackets for their stepper motor, so I determined to modify their bracket to hold the replacement stepper motor. The same suppliers were also able to supply a 50W tungsten-halogen lamp complete with lamp housing, which would be the ideal light source for the wavelength range of interest. While this was ludicrously expensive for what was undoubtedly a slide projector bulb, it had the advantage of a ready-made housing complete with a lens to form a collimated beam. Unfortunately the lamp housing would be too large to fit inside the instrument case, but it could be bolted to the rear of the case. (The "case" in this instance would be fitted into a much larger cabinet holding the complete system, so this did not really present any difficulties. In fact it was a distinct advantage, in that it both left more room inside the unit's

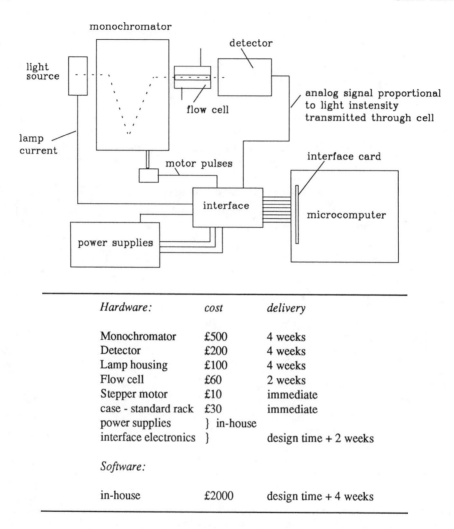

Fig 10.2 *Principal components required for proposed variable wavelength colorimeter.*

Hardware:	cost	delivery
Monochromator	£500	4 weeks
Detector	£200	4 weeks
Lamp housing	£100	4 weeks
Flow cell	£60	2 weeks
Stepper motor	£10	immediate
case - standard rack	£30	immediate
power supplies	} in-house	
interface electronics	}	design time + 2 weeks
Software:		
in-house	£2000	design time + 4 weeks

case, and removed a significant source of heat from the environment of the sensitive measuring electronics.)

A small silicon photodiode with a built-in amplifier was available from the same manufacturer and would provide an ideal light-detecting transducer, with the added advantage that the integrated amplifier would minimise the chance of interference pick-up from the light source's power supply (which would have to be located with the case).

So, some of the major components were available commercially, at an established price and available on short delivery. The other components required would be the computer interface and the power supplies: 12 V, 50 W for the lamp; 12 V, 100 mA for the stepper

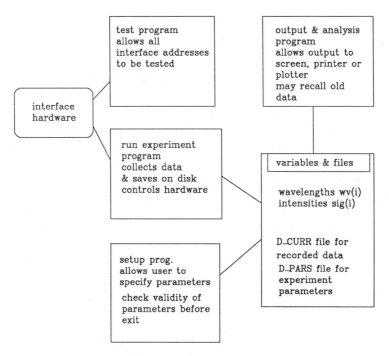

Fig 10.3 *Software overview.*

motor; ± 12 V for the analog circuitry - including the amplifier integrated into the photodiode assembly (say 50 mA); and 5 V for the logic system associated with the interface and stepper motor logic (say 100 mA). These could be designed and constructed in-house.

At this stage a sketch of the proposed system was drawn and is reproduced in fig 10.2, along with the major components list.

Step 6. Devise software overview

The software for operating the unit could be relatively simple in design because user interactions could be quite separate from the actual operation of the detector. Thus a subprogram which allowed the user to specify the values of required parameters (SETUP) would be designed to present to the user the same general appearance as the other components of the software already in use on the analyser system. The subprogram which would operate the monochromator and record the intensity measurements (RUNNER) would need no direct user interaction apart from premature termination - indeed it was to be expected that the system would normally operate when no user was present - although results would be presented on the screen as they were recorded to simplify testing and monitoring. A third subprogram (REPORTER) would enable results to be viewed on the screen or printed on a printer on demand. A fourth subprogram (TESTER) would remain hidden from the normal user, and would only be accessible via a password (because it would allow the operator to turn on the lamp for extended periods for testing purposes). With one eye on the most likely future enhancement, provision would be made within USER to enable the user to specify a spectral scan of each sample (a facility which in was in any case required within TESTER in order to determine the no-sample intensity at each wavelength to check the condition of the lamp). REPORTER would also be equipped with facilities to provide spectral output (using the same routines as TESTER), although these spectral facilities would not appear on selection menus of the user's software until the upgrade was requested. The required file structure would consist of the user specified parameters (the wavelength, and the rate of sampling), a few spare records for future expansion, then an unspecified number of records containing the date, time, sample number, and peak height for each sample recorded. The sampling rate which could be anticipated for the system was such that a maximum allowable number of records of 1000 in any one file was expected to cope with any foreseeable demand. A file structure suitable for spectral data was also defined, although in the first version of the software the data involved was not recorded on disk and was only used for screen and printer presentation for lamp checking purposes.

I did not anticipate any software difficulties in achieving the priorities assigned to sensitivity or cell volume. The priority specified for continuous operation and security of recorded data led me to decide on a write to disk after every reading followed by a closure of the disk file. This had the advantage of allowing recorded data to remain intact through a power failure. I considered the possibility of allowing the system to automatically resume after a power failure, but this was thought undesirable for reasons unconnected with the detector unit. The ease of use priority presented no problems with this system, as simple menu driven software could easily cope with the limited range of functions required for the user's interaction with the system. Finally the timescales were such that the system could be handled from BASIC, with machine code drivers for the interface functions. I decided that I would use QuickBASIC because I prefer it.

The software overview chart which resulted from these thoughts is shown in fig 10.3.

Step 7. Outline design/specify hardware components

An examination of the system components in the light of the operational concept and the system specification indicated that the signals involved in measurement and control functions were as follows:

a) a stable high current supply to power the light source.
b) output signals to operate the 12 V stepper motor.
c) input logic levels to indicate when the monochromator grating had reached the upper or lower limits of its travel.
d) an input signal proportional to the light intensity.

These considerations lead to an outline design shown in fig 10.4.

The stepper motor stepping angle could be 7.5° as this corresponded to a wavelength change of only 0.5 nm according to the monochromator specification provided by the

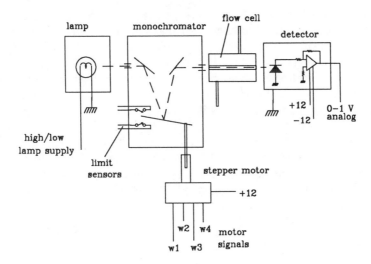

Fig 10.4 *Outline design showing hardware components.*

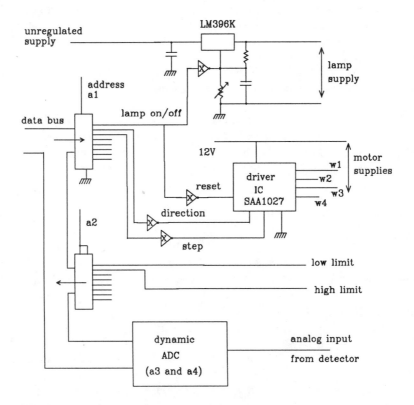

Fig 10.5 *Outline design of stepper motor & lamp interface.*
x = 5/12 V logic level changes.

manufacturer. Thus 800 steps would cover the entire wavelength range required, and the specifications called for this range to be covered in a minimum of 20 seconds. As there should be no difficulty in producing the required 40 steps per second by software, it seemed reasonable to allow a single bit of an output byte to be used for signalling "step" to the stepper motor driving electronics. The stepper motor actually requires several pulsed currents to produce a single step, but as these are only about 100 mA a simple IC stepper motor controller would be able to drive the motor directly. An additional bit of the same output byte could be employed to set the step direction. It had already been noted that the light source should be under some control from the computer, although the requirement was really to turn the source from its normal operating power level to a standby level (which would keep the bulb warm while prolonging its lifespan). An additional bit of the output byte being used for stepper motor control could be used to provide a high/low signal for the lamp. The monochromator could be fitted with light emitting diodes and photodiodes to produce a logic level change when the grating reached either limit of its travel, so that two bits of an input byte would be required to sense an "end of travel" condition.

The high performance silicon photodiode with integral amplifier could generate an output voltage within the range 0-10 V when detecting the signal from a 50 W quartz-halogen lamp through the monochromator (information in manufacturer's brochure), so I assumed that it should be able to manage 0-1 V over the wavelength range of interest. To read this signal into the computer with the specified precision (0.02%-0.05%) would require an analog-to-digital converter with 12 or 13 bits of digital output. For reasons of simplicity and low cost, and because I was already using several dynamic analog input interfaces in other systems, I opted for a dynamic interface of the kind discussed in section 7.7.

The outline diagram of the interface system suggested by these thoughts is shown in fig 10.5. The major contributions to the cost of this system could be determined from the prices of the obvious ICs, power supply components and circuit board, with a small amount added for the cost of the additional (but low cost) logic and passive components which would undoubtedly be required.

Step 8. Decide whether to proceed

There seemed no reason to suppose that building the instrument would present any major difficulties, and the total cost estimated so far was under the target figure. I estimated that the instrument could be completed in one month. The company paying the bills approved.

Step 9. Design main software flowchart

Given the software overview chart and the information outlined above about the interface characteristics, the elements of the main software flowchart were readily assembled. The main flow chart devised was too large to reproduce here, so fig 10.6 is limited to two portions of the chart concerned with the main program and the subprogram RUNNER. The main program merely calls routines to handle initial power-up testing of the hardware, checks that it is all in a safe state, and then displays the main user menu and current status (and the logo - all programs should have a friendly logo to reassure the user that all is well. In our

a)

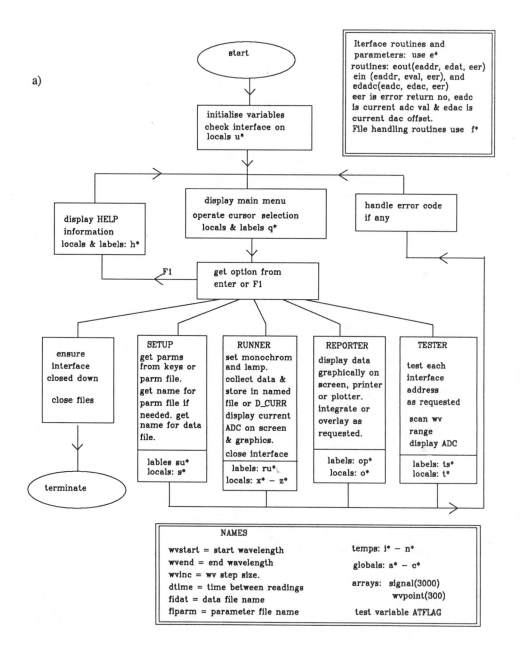

Fig 10.6 *Sections of the main flowchart showing the pages for (a) the main program and (b) subprogram RUNNER.*

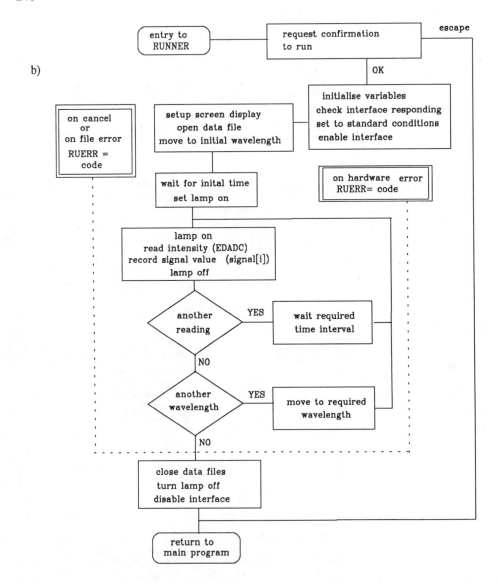

b)

systems the logo generally includes a status line which indicates what the system is currently doing. At this point it is idle.). The user is prompted to choose the activity required using only the cursor keys and <enter>. No other keys do anything except cause a beep to indicate "wrong key". The activities offered are limited to four choices, so relatively little study of the screen is required. One choice (collect data) leads to a call to RUNNER, which is the only subprogram which operates the interface to the detector.

On entry to RUNNER the program checks that some parameters exist (ie that SETUP has been used since the power was turned on), that the parameters available are sensible (ie

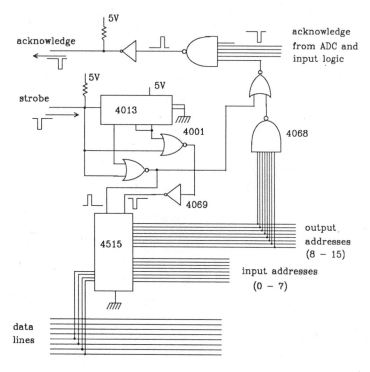

Fig 10.7 *Circuit diagram of the address decoder section of the interface (pin numbers have been omitted for clarity).*

wavelength limits are within the system's range, etc.), and that enough disk storage space exists for the required data storage. If any problem is encountered RUNNER returns control to the main program with a suitable error code so that the main program can display a helpful error message in a standard format. If all is well RUNNER checks that the interface power has been on for at least five minutes (to ensure that the power supplies and light source are "warm"), then sets the monochromator to the desired initial wavelength (as specified in SETUP) and starts recording data according to the predefined parameters. The data is displayed as it is recorded and the appropriate sample numbers, times, dates and peak heights stored on disk. When the specified number of samples have been recorded the current status is amended and control is returned to the main program.

The main flowchart indicates the actions which need to be taken by the routines called and the access which such routines may require to files and variables. However, it does not at this stage need to supply detailed information, such as the addresses which routines will use to communicate with specific interface functions. In fact, such addresses may not be finalised until the hardware design is completed (a particular address distribution may be desirable on, say, a printed circuit card in order to minimise the complexity of the track layout), or until the detailed software flowcharts have been completed (where, for example, addresses specified for commercial interface units are adopted).

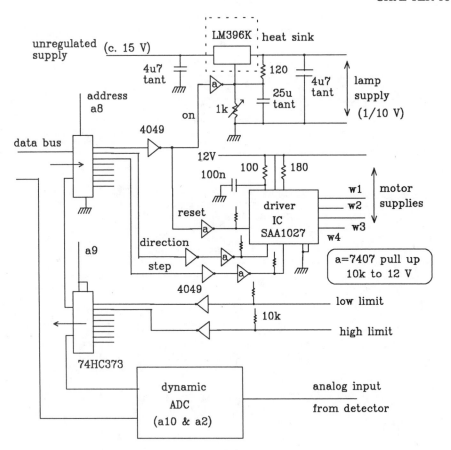

Fig 10.8 *Circuit diagram of the stepper motor and lamp controller section*
of the interface (pin numbers have been omitted for clarity).

Step 10. Design in-house hardware

The requirements for in-house hardware components have been laid down at step 7, and
the interaction between the hardware and the software had been outlined (although not
detailed) in step 9. The need now was to detail the mechanical and circuit diagrams of the
hardware components to be assembled in-house, and define any interface parameters (eg.
interface addresses, byte and control line activities) necessary before the software
requirements could be specified The circuit diagrams for the two elements of the interface
system are shown in figs 10.7 and 10.8, the address decoder circuitry and the stepper motor/
lamp supply control circuitry respectively. The operations of these circuits have been
described in chapter 7. Circuit diagrams were also produced for the dynamic analog input
circuitry (see chapter 7) and for the remaining power supplies.

These circuits defined the sequence of operations required for communication between
the micro and the interface, including the interface addresses as illustrated in fig 10.8.

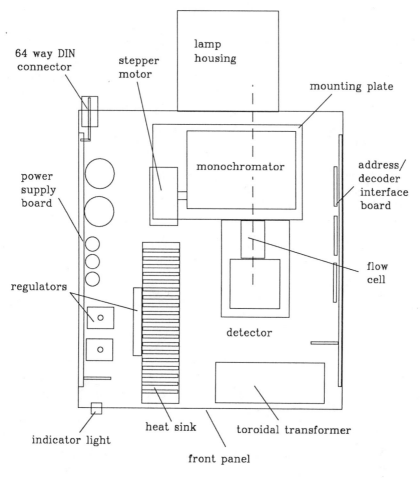

Fig 10.9 *Layout of completed colorimetric detector.*

The unit was constructed on two Eurocard boards. The power supplies (less the transformer) were put on a stripboard, and the interface circuitry on a wire-wrap board. Sockets were used for all ICs. The circuit cards were mounted at the sides of the case, with the transformer, monochromator, light source and detector mounted on the base plate. A diagram of the completed assembly is shown in fig 10.9. Connection to the computer (and to other components of the overall system) was made via a 64-way connector at the rear of the case (the case was a plug-in module which was housed in a larger cabinet).

Step 11. Draw software flowcharts

From the main software flowchart of step 9, details of system components (eg. the requirements of screen graphics, printer and plotter), and the sequence of operations required for communication with the interface, the detailed flowcharts for the individual routines and

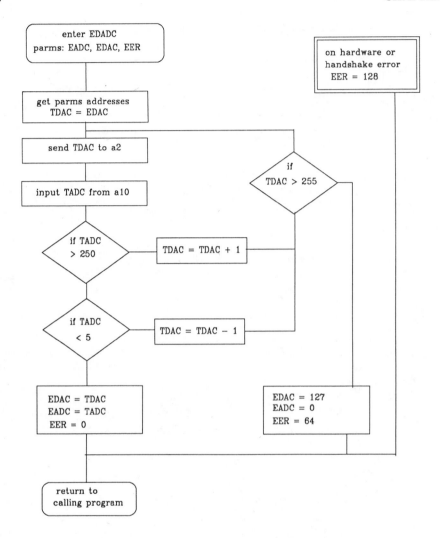

Fig 10.10 *Detailed flowchart for EDADC routine.*

functions of the software could be drawn out. A single routine flowchart, that for the reading of the dynamic analog input interface - routine EDADC, is reproduced in fig 10.10. It should be noted that this routine does call two other routines (one for byte input and one for byte output), and that the names of routines and variables (at least the first letters of the names) have already been specified in step 6.

As EDADC would be required to return a result within a few milliseconds, the flowchart was annotated to indicate that the routine should be coded in assembler language. Note the presence of the test on entry to determine whether interface communication should take place or whether random results should be returned for test purposes.

Step 12. Write software

Using the many individual flowcharts drawn up at step 11 it was a relatively straightforward matter to code the routines in the chosen languages (QuickBASIC and assembler), and to compile or assemble the individual files of code. The assembler version of EDADC is shown in the listing below. Routines were grouped into files entirely on the basis of programmer convenience, and the only testing carried out at this stage was to ensure that the software would correctly handle the screen and disk files. The ability to run the software without interface hardware connected (using tests of the kind outlined above) is invaluable at this stage.

Step 13. Test components and assemble system

Testing was carried out systematically. The power supply levels were checked before the ICs were inserted (this is always much cheaper than replacing all ICs when a power supply fault is discovered the hard way). The address decoder circuit ICs were inserted and a BASIC program used to send addresses and data bytes to the interface. An oscilloscope was used to ensure that the address lines responded correctly. The ADC/DAC components were then inserted and the BASIC test program used to ensure that the DAC produced a suitable output. A simulated analog input signal (a variable resistor and a 1.5 V battery) was used to enable the ADC to be tested, then a BASIC version of EDADC was used to check the (slow) operation of the dynamic analog input.

The stepper motor controller and logic ICs were inserted and a BASIC routine used to check that the stepper motor moved correctly and that the limit could be sensed. The light source lamp was inserted and the lamp on/off line tested. Unfortunately the LM396K regulator was getting too hot. A larger heat sink was installed and bolted inside the front panel of the case.

Calibration of the monochromator wavelength was checked using borrowed filters placed in the light path. If we assumed that the lower limit switch operated at 10 nm then the calibration over the range 400 - 700 nm was within the manufacturer's specification. (The software had been written on the assumption that the lower limit switch operated at 0 nm. A recompilation was required.)

The system was then fully assembled and tested with the real software. It worked. For the next month the software (ie the development version) was modified a number of times to remove minor bugs and inconveniences. The most tedious aspect was ensuring that the flowcharts were modified to correctly define the software. However, the instrument performed well up to expectations and the supporting company was well pleased.

```
:       Microsoft assembler coded 4:1:86          ifin:   push    bp       : put BP on stack
code    SEGMENT      public 'code'                        mov     bp,sp     :  then  collect  parameter
        assume cs:code, ds:code                   addresses
main    proc far        :  ref fiastar 1.0                call    pair
        public  main                                      mov     bx,ado
        public  EDOUT                                     mov     al,[bx]
edout:  jmp     ifout                                     mov     adr,al
        public  EDIN                                      call    inb
edin:   jmp     ifin                                      mov     bx,dao
        public  EDADC                                     mov     al,dat
edadc:  jmp     anal                                      mov     [bx],al
                                                          jmp     done
max     equ     245      : max value of ADC reading
min     equ     10       : min value of ADC reading  done:   mov     al,3     : when finished set PB high
adca    equ     2        : address of ADC                  mov     dx,pb    : put 255 in port A
daca    equ     10       : address of DAC                  out     dx,al
pa      equ     1000     : interface adaptor port A        mov     al,255
pb      equ     1001     : interface adaptor port B        mov     dx,pa
pc      equ     1002     : interface adaptor port C        out     dx,al
fl      db      ?        : flag                            mov     bx,ero   : store current error code
adr     db      0        : current address value          mov     al,erv   : in error parameter
dat     db      ?        : current data byte value         mov     [bx],al
erv     db      0        : current error code              pop     bp
ado     dw      ?        : address value of ADC param      ret     6        : return to calling program
dao     dw      ?        : address value of DAC param  main    endp
ero     dw      ?        : address value of error param
dac     db      127      : last DAC byte value         pair    proc    near
cou     db      0        : counter for timing waits           mov     bx,8[bp] : get a three param addresses
                                                          mov     dao,bx          : from stack.
anal:   push    bp       : put BP on stack                mov     bx,10[bp]
        mov     bp,sp    : then collect param addresses    mov     ado,bx
        call    pair                                      mov     bx,6[bp]
        call    analog   : read dynamic analog input       mov     ero,bx
        mov     bx,ado                                    mov     al,0
        mov     al,dat                                    mov     erv,al
        mov     [bx],al                                   mov     cou,al   : zero counter and error code
        mov     bx,dao                                    ret
        mov     al,dac                            pair    endp
        mov     [bx],al
        jmp     done                              adrs    proc    near     : transmit address from adr
                                                          mov     al,128d  : to interface
ifout:  push    bp       : put BP on stack                mov     fl,al
        mov     bp,sp    : then collect param addresses    mov     al,3
        call    pair                                      mov     dx,pb
        mov     bx,ado                                    out     dx,al    : strobe hi
        mov     al,[bx]                                   mov     al,adr
        mov     adr,al                                    mov     dx,pa
        mov     bx,dao                                    out     dx,al
        mov     al,[bx]                                   mov     al,2
        mov     dat,al                                    mov     dx,pb
        call    outb                                      out     dx,al    : strobe lo
        jmp     done
```

Listing of assembler language interface handling routines incoporating EDADC

```
        call    wflag    : wait for acknowledge              jz      tup     : junp if time exceeded
        mov     al,3                                         mov     cx,255
        mov     dx,pb                                        jmp     test    : check again
        out     dx,al    : strobe hi              tup:       mov     al,128  : if time up error code = 128
        ret                                                  mov     erv,al
adrs    endp                                      ok:        ret
                                                  wflag      endp
outb    proc    near
        call    adrs     : call output address    analog proc  near        : dynamic analog input routine
        mov     al,dat   : (leaves strobe hi)                 jmp     odac    : send out current dac value
        mov     dx,pa    : then output a byte from dat read:  mov     al,adca : rem read adc valu
        out     dx,al                                        mov     adr,al
        mov     dx,pb    : strobe lo                          call    inb
        mov     al,2                                         mov     al,erv
        out     dx,al                                        cmp     al,0    : check no error
        call    wflag    : wait for acknowledge              jne     vals
        mov     dx,pb                                        mov     al,dat
        mov     al,3                                         cmp     al,max  : compare with max
        out     dx,al    : strobe hi                         ja      tosmal
        ret                                                  cmp     al,min  : and min values allowed
outb    endp                                                 jb      tobig
                                                  vals:      ret
inb     proc    near                              tobig:     mov     al,daca : if adc>max then inc dac
        call    adrs     : call output address               mov     adr,al  : repeat unless dac > 255
                                                             mov     al,dac
        mov     al,32d   : (leaves strobe hi)                cmp     al,255
        mov     fl,al    : change ack mask  in fl to 32      je      fail
        mov     al,5                                         inc     al
        mov     dx,pb    : strobe hi                         mov     dac,al
        out     dx,al                                        jmp     odac
        mov     al,4                              odac:      mov     al,daca : send current dac value
        out     dx,al    : strobe lo                         mov     adr,al
        call    wflag    : wait for acknowledge              mov     al,dac
        mov     dx,pa                                        mov     dat,al
        in      al,dx    : input a byte from port A          call    outb
        mov     dat,al                                       mov     cx,255
        mov     al,5                              tosmal: mov al,dac  : if adc<min the dec dac
        mov     dx,pb                                        cmp     al,1d   : repeat unless dac<1
        out     dx,al    : strobe hi                         jb      fail
        mov     al,3                                         dec     al
        out     dx,al    : set ack for output reply          mov     dac,al
        ret                                                  jmp     odac
inb     endp                                      fail:      mov     al,127  : DAC set = 127 if value
                                                             mov     dac,al  : out of range.
wflag   proc    near     : wait for acknowledge routine      mov     al,64   : error code =64
        mov     cx,255   : set inner loop index to 255       mov     erv,al
        mov     dx,pc                                        jmp     vals
test:   in      al,dx    : input port C          blurb      db      40 dup(?)        : spare bytes
        and     al,fl    : and with fl mask      analog     endp
        jnz     ok       : non-zero if ack received code     ends
        loop    test     : else wait
        dec     cou      : cou holds outer loop counter      end     main
```

APPENDIX 1 The American Standard Code for Information Interchange
(The ASCII code)

dec	hex	character	name
0	00	NUL	null character
1	01	SOH	start of heading
2	02	STX	start text
3	03	ETX	end text
4	04	EOT	end of transmission
5	05	ENQ	enquiry
6	06	ACK	acknowledge
7	07	BEL	ring bell
8	08	BS	backspace
9	09	HT	horizontal tab
10	0a	LF	linefeed
11	0b	VT	vertical tab
12	0c	FF	formfeed
13	0d	CR	carriage return
14	0e	SO	shift out
15	0f	SI	shift in
16	10	DLE	data link escape
17	11	DC1	device control #1
18	12	DC2	device control #2
19	13	DC1	device control #3
20	14	DC1	device control #4
21	15	NAK	negative acknowledge
22	16	SYN	synchronous idle
23	17	ETB	end transmission block
24	18	CAN	cancel
25	19	EM	end medium
26	1a	SUB	substitute
27	1b	ESC	escape
28	1c	FS	file separator
29	1d	GS	group separator
30	1e	RS	record separator
31	1f	US	unit separator
32	20	SP	space

The ASCII code (contd.)

dec	hex	char	dec	hex	char	dec	hex	char
33	21	!	65	41	A	97	61	a
34	22	"	66	42	B	98	62	b
35	23	#	67	43	C	99	63	c
36	24	$	68	44	D	100	64	d
37	25	%	69	45	E	101	65	e
38	26	&	70	46	F	102	66	f
39	27	'	71	47	G	103	67	g
40	28	(	72	48	H	104	68	h
41	29	)	73	49	I	105	69	i
42	2a	*	74	4a	J	106	6a	j
43	2b	+	75	4b	K	107	6b	k
44	2c	,	76	4c	L	108	6c	l
45	2d	-	77	4d	M	109	6d	m
46	2e	.	78	4e	N	110	6e	n
47	2f	/	79	4f	O	111	6f	o
48	30	0	80	50	P	112	70	p
49	31	1	81	51	Q	113	71	q
50	32	2	82	52	R	114	72	r
51	33	3	83	53	S	115	73	s
52	34	4	84	54	T	116	74	t
53	35	5	85	55	U	117	75	u
54	36	6	86	56	V	118	76	v
55	37	7	87	57	W	119	77	w
56	38	8	88	58	X	120	78	x
57	39	9	89	59	Y	121	79	y
58	3a	:	90	5a	Z	122	7a	z
59	3b	;	91	5b	[	123	7b	{
60	3c	<	92	5c	\	124	7c	|
61	3d	=	93	5d	]	125	7d	}
62	3e	>	94	5e	^	126	7e	~
63	3f	?	95	5f	_	127	7f	del
64	40	@	94	60	'			

APPENDIX 2 Binary and Binary Coded Decimal (BCD) values

binary	dec	BCD	Binary	dec	BCD	binary	dec	BCD
00000001	1	01	00000010	2	02	00000011	3	03
00000100	4	04	00000101	5	05	00000110	6	06
00000111	7	07	00001000	8	08	00001001	9	09
00001010	10	—	00001011	11	—	00001100	12	—
00001101	13	—	00001110	14	—	00001111	15	—
00010000	16	10	00010001	17	11	00010010	18	12
00010011	19	13	00010100	20	14	00010101	21	15
00010110	22	16	00010111	23	17	00011000	24	18
00011001	25	19	00011010	26	—	00011011	27	—
00011100	28	—	00011101	29	—	00011110	30	—
00011111	31	—	00100000	32	20	00100001	33	21
00100010	34	22	00100011	35	23	00100100	36	24
00100101	37	25	00100110	38	26	00100111	39	27
00101000	40	28	00101001	41	29	00101010	42	—
00101011	43	—	00101100	44	—	00101101	45	—
00101110	46	—	00101111	47	—	00110000	48	30
00110001	49	31	00110010	50	32	00110011	51	33
00110100	52	34	00110101	53	35	00110110	54	36
00110111	55	37	00111000	56	38	00111001	57	39
00111010	58	—	00111011	59	—	00111100	60	—
00111101	61	—	00111110	62	—	00111111	63	—
01000000	64	40	01000001	65	41	01000010	66	42
01000011	67	43	01000100	68	44	01000101	69	45
01000110	70	46	01000111	71	47	01001000	72	48
01001001	73	49	01001010	74	—	01001011	75	—
01001100	76	—	01001101	77	—	01001110	78	—
01001111	79	—	01010000	80	50	01010001	81	51
01010010	82	52	01010011	83	53	01010100	84	54
01010101	85	55	01010110	86	56	01010111	87	57
01011000	88	58	01011001	89	59	01011010	90	—
01011011	91	—	01011100	92	—	01011101	93	—
01011110	94	—	01011111	95	—	01100000	96	60
01100001	97	61	01100010	98	62	01100011	99	63
01100100	100	64	01100101	101	65	01100110	102	66
01100111	103	67	01101000	104	68	01101001	105	69
01101010	106	—	01101011	107	—	01101100	108	—
01101101	109	—	01101110	110	—	01101111	111	—
01110000	112	70	01110001	113	71	01110010	114	72
01110011	115	73	01110100	116	74	01110101	117	75
01110110	118	76	01110111	119	77	01111000	120	78
01111001	121	79	01111010	122	—	01111011	123	—
01111100	124	—	01111101	125	—	01111110	126	—
01111111	127	—	10000000	128	80	10000001	129	81
10000010	130	82	10000011	131	83	10000100	132	84
10000101	133	85	10000110	134	86	10000111	135	87
10001000	136	88	10001001	137	89	10001010	138	—
10001011	139	—	10001100	140	—	10001101	141	—
10001110	142	—	10001111	143	—	10010000	144	90
10010001	145	91	10010010	146	92	10010011	147	93
10010100	148	94	10010101	149	95	10010110	150	96
10010111	151	97	10011000	152	98	10011001	153	99
10011010	154	—	10011011	155	—	10011100	156	—
10011101	157	—	10011110	158	—	10011111	159	—

Note: — represents a forbidden BCD representation.

APPENDIX 3. Decimal-Hexadecimal conversion tables

a) 1-256 in steps of 1

dec	hex	dec	hex	dec	hex	dec	hex	dec	hex	dec	hex
1	01H	2	02H	3	03H	4	04H	5	05H	6	06H
7	07H	8	08H	9	09H	10	0aH	11	0bH	12	0cH
13	0dH	14	0eH	15	0fH	16	10H	17	11H	18	12H
19	13H	20	14H	21	15H	22	16H	23	17H	24	18H
25	19H	26	1aH	27	1bH	28	1cH	29	1dH	30	1eH
31	1fH	32	20H	33	21H	34	22H	35	23H	36	24H
37	25H	38	26H	39	27H	40	28H	41	29H	42	2aH
43	2bH	44	2cH	45	2dH	46	2eH	47	2fH	48	30H
49	31H	50	32H	51	33H	52	34H	53	35H	54	36H
55	37H	56	38H	57	39H	58	3aH	59	3bH	60	3cH
61	3dH	62	3eH	63	3fH	64	40H	65	41H	66	42H
67	43H	68	44H	69	45H	70	46H	71	47H	72	48H
73	49H	74	4aH	75	4bH	76	4cH	77	4dH	78	4eH
79	4fH	80	50H	81	51H	82	52H	83	53H	84	54H
85	55H	86	56H	87	57H	88	58H	89	59H	90	5aH
91	5bH	92	5cH	93	5dH	94	5eH	95	5fH	96	60H
97	61H	98	62H	99	63H	100	64H	101	65H	102	66H
103	67H	104	68H	105	69H	106	6aH	107	6bH	108	6cH
109	6dH	110	6eH	111	6fH	112	70H	113	71H	114	72H
115	73H	116	74H	117	75H	118	76H	119	77H	120	78H
121	79H	122	7aH	123	7bH	124	7cH	125	7dH	126	7eH
127	7fH	128	80H	129	81H	130	82H	131	83H	132	84H
133	85H	134	86H	135	87H	136	88H	137	89H	138	8aH
139	8bH	140	8cH	141	8dH	142	8eH	143	8fH	144	90H
145	91H	146	92H	147	93H	148	94H	149	95H	150	96H
151	97H	152	98H	153	99H	154	9aH	155	9bH	156	9cH
157	9dH	158	9eH	159	9fH	160	a0H	161	a1H	162	a2H
163	a3H	164	a4H	165	a5H	166	a6H	167	a7H	168	a8H
169	a9H	170	aaH	171	abH	172	acH	173	adH	174	aeH
175	afH	176	b0H	177	b1H	178	b2H	179	b3H	180	b4H
181	b5H	182	b6H	183	b7H	184	b8H	185	b9H	186	baH
187	bbH	188	bcH	189	bdH	190	beH	191	bfH	192	c0H
193	c1H	194	c2H	195	c3H	196	c4H	197	c5H	198	c6H
199	c7H	200	c8H	201	c9H	202	caH	203	cbH	204	ccH
205	cdH	206	ceH	207	cfH	208	d0H	209	d1H	210	d2H
211	d3H	212	d4H	213	d5H	214	d6H	215	d7H	216	d8H
217	d9H	218	daH	219	dbH	220	dcH	221	ddH	222	deH
223	dfH	224	e0H	225	e1H	226	e2H	227	e3H	228	e4H
229	e5H	230	e6H	231	e7H	232	e8H	233	e9H	234	eaH
235	ebH	236	ecH	237	edH	238	eeH	239	efH	240	f0H
241	f1H	242	f2H	243	f3H	244	f4H	245	f5H	246	f6H
247	f7H	248	f8H	249	f9H	250	faH	251	fbH	252	fcH
253	fdH	254	feH	255	ffH	256	100H				

b) 256-32768 in steps of 256

dec	hex	dec	hex	dec	hex	dec	hex	dec	hex
256	100H	512	200H	768	300H	1024	400H	1280	500H
1536	600H	1792	700H	2048	800H	2304	900H	2560	a00H
2816	b00H	3072	c00H	3328	d00H	3584	e00H	3840	f00H
4096	1000H	4352	1100H	4608	1200H	4864	1300H	5120	1400H
5376	1500H	5632	1600H	5888	1700H	6144	1800H	6400	1900H
6656	1a00H	6912	1b00H	7168	1c00H	7424	1d00H	7680	1e00H
7936	1f00H	8192	2000H	8448	2100H	8704	2200H	8960	2300H
9216	2400H	9472	2500H	9728	2600H	9984	2700H	10240	2800H
10496	2900H	10752	2a00H	11008	2b00H	11264	2c00H	11520	2d00H
11776	2e00H	12032	2f00H	12288	3000H	12544	3100H	12800	3200H
13056	3300H	13312	3400H	13568	3500H	13824	3600H	14080	3700H
14336	3800H	14592	3900H	14848	3a00H	15104	3b00H	15360	3c00H
15616	3d00H	15872	3e00H	16128	3f00H	16384	4000H	16640	4100H
16896	4200H	17152	4300H	17408	4400H	17664	4500H	17920	4600H
18176	4700H	18432	4800H	18688	4900H	18944	4a00H	19200	4b00H
19456	4c00H	19712	4d00H	19968	4e00H	20224	4f00H	20480	5000H
20736	5100H	20992	5200H	21248	5300H	21504	5400H	21760	5500H
22016	5600H	22272	5700H	22528	5800H	22784	5900H	23040	5a00H
23296	5b00H	23552	5c00H	23808	5d00H	24064	5e00H	24320	5f00H
24576	6000H	24832	6100H	25088	6200H	25344	6300H	25600	6400H
25856	6500H	26112	6600H	26368	6700H	26624	6800H	26880	6900H
27136	6a00H	27392	6b00H	27648	6c00H	27904	6d00H	28160	6e00H
28416	6f00H	28672	7000H	28928	7100H	29184	7200H	29440	7300H
29696	7400H	29952	7500H	30208	7600H	30464	7700H	30720	7800H
30976	7900H	31232	7a00H	31488	7b00H	31744	7c00H	32000	7d00H
32256	7e00H	32512	7f00H	32768	8000H						

c) 33024-65536 in steps of 512

dec	hex	dec	hex	dec	hex	dec	hex	dec	hex
33024	8100H	33536	8300H	34048	8500H	34560	8700H	35072	8900H
35584	8b00H	36096	8d00H	36608	8f00H	37120	9100H	37632	9300H
38144	9500H	38656	9700H	39168	9900H	39680	9b00H	40192	9d00H
40704	9f00H	41216	a100H	41728	a300H	42240	a500H	42752	a700H
43264	a900H	43776	ab00H	44288	ad00H	44800	af00H	45312	b100H
45824	b300H	46336	b500H	46848	b700H	47360	b900H	47872	bb00H
48384	bd00H	48896	bf00H	49408	c100H	49920	c300H	50432	c500H
50944	c700H	51456	c900H	51968	cb00H	52480	cd00H	52992	cf00H
53504	d100H	54016	d300H	54528	d500H	55040	d700H	55552	d900H
56064	db00H	56576	dd00H	57088	df00H	57600	e100H	58112	e300H
58624	e500H	59136	e700H	59648	e900H	60160	eb00H	60672	ed00H
61184	ef00H	61696	f100H	62208	f300H	62720	f500H	63232	f700H
63744	f900H	64256	fb00H	64768	fd00H	65280	ff00H		

d) 65536-4194304 in steps of 65536

dec	hex	dec	hex	dec	hex	dec	hex
65536	10000H	131072	20000H	196608	30000H	262144	40000H
327680	50000H	393216	60000H	458752	70000H	524288	80000H
589824	90000H	655360	a0000H	720896	b0000H	786432	c0000H
851968	d0000H	917504	e0000H	983040	f0000H	1048576	100000H
1114112	110000H	1179648	120000H	1245184	130000H	1310720	140000H
1376256	150000H	1441792	160000H	1507328	170000H	1572864	180000H
1638400	190000H	1703936	1a0000H	1769472	1b0000H	1835008	1c0000H
1900544	1d0000H	1966080	1e0000H	2031616	1f0000H	2097152	200000H
2162688	210000H	2228224	220000H	2293760	230000H	2359296	240000H
2424832	250000H	2490368	260000H	2555904	270000H	2621440	280000H
2686976	290000H	2752512	2a0000H	2818048	2b0000H	2883584	2c0000H
2949120	2d0000H	3014656	2e0000H	3080192	2f0000H	3145728	300000H
3211264	310000H	3276800	320000H	3342336	330000H	3407872	340000H
3473408	350000H	3538944	360000H	3604480	370000H	3670016	380000H
3735552	390000H	3801088	3a0000H	3866624	3b0000H	3932160	3c0000H
3997696	3d0000H	4063232	3e0000H	4128768	3f0000H	4194304	400000H

BIBLIOGRAPHY

General electronics and integrated circuits

O. Bishop, "Beginner's Guide to Electronics", 4th edition
pub. Newnes, London, UK., 1982

G. B. Clayton, "Operational Amplifiers", 2nd edition
pub. Butterworths, London, UK., 1979

J. K. Hardy, "High frequency circuit design"
pub. Reston Publishing Co. Inc., Reston, Virginia, 1979

B. Holdsworth, "Digital Logic Design", 2nd edition
pub. Butterworths, London, UK., 1987

P. Horowitz and W. Hill, "The Art of Electronics"
pub. Cambridge University Press, Cambridge, UK., 1980

E. Hnatek, "Design of Solid State Power Supplies", 2nd. edition
pub. Van Nostrand Reinhold, New York, NY., 1981

D. Johnson, L. Johnson, and H. Moore, "A Handbook of Active Filters"
pub. Prentice-Hall, Englewood Cliffs, NJ., 1980

W. G. Jung, "IC Op-amp Cookbook", 3rd edition
pub. Howard W. Sams & Co. Inc, Indianapolis, Indiana, 1986

R. Morrison, "Grounding and Shielding Techniques in Instrumentation", 3rd. edition
pub. J. Wiley & Sons, New York, NY., 1986

E. S. Oxner, "Power FETs and their Applications"
pub. Prentice-Hall, Englewood Cliffs, NJ., 1982

R. G. Seippel, "Transducers, Sensors and Detectors"
pub. Prentice-Hall, Englewood Cliffs, NJ., 1983

Microcomputers and software

R. Duncan, "Advanced MS-DOS"
    pub. Microsoft Press, Bellevue, Washington, 1986

D. Durant, G. Carlson and P. Yao, "Programmer's Guide to Windows"
    pub. Sybex Inc., Alameda, California, 1987

S. Dvorak and A. Musset, "BASIC in Action"
    pub. Butterworths, London, UK., 1984

M. H. Lewin, "Elements of C"
    pub. Plenum Publishing, New York, NY, 1986

C. Morgan and M. Waite, "8086/8088 16-bit Microprocessor Primer"
    pub. McGraw-Hill, Berkeley, CA., 1982

P. Norton, "Programmer's guide to the IBM PC"
    pub. Microsoft Press, Bellevue, Washington, 1985

S. P. Morse, "The 8086/8088 Primer"
    pub. Hayden, London, UK., 1982

M. Sargent & R. L. Shoemaker, "The IBM Personal Computer from the inside out"
    pub. Addison-Wesley Publishing Co. Inc. Reading, MA., 1984

L. J. Scanlon, "IBM PC & XT Assembly Language"
    pub. Brady Inc, New York, NY., 1985

R. Startz, "8087 Applications and Programming"
    pub. R. J. Brady Co., Bowie, MD, 1983

R. J. Traister, "Going from BASIC to C"
    pub. Prentice-Hall, Englewood Cliffs, NJ., 1985

The Waite Group, "68000, 68010, 68020 Primer"
    pub. Howard W. Sams & Co. Inc, Indianapolis, Indiana, 1985

V. Wolverton, "Running MS-DOS"
    pub. Microsoft Press, Bellevue, Washington, 1984

Interfacing to digital systems

B. A. Artwick, "Microcomputer Interfacing"
    pub. Prentice-Hall, Englewood Cliffs, NJ., 1980

J. C. Cluley, "Interfacing to Microprocessors"
    pub. Macmillan Publishing, Basingstoke, UK., 1983

P. H. Garrett, "Analog I/O Design"
    pub. Prentice-Hall, Englewood Cliffs, NJ., 1981

P. W. Gofton, "Mastering Serial Communications"
    pub. Sybex Inc., Alameda, California, 1987

C. Pye, "Networking with Microcomputers"
    pub. NCC Publications, Manchester, UK., 1985

Martin D. Seyer, "RS-232 Made Easy"
    pub. Prentice-Hall, Englewood Cliffs, NJ., 1984

D. A. Tugal and O. Tugal, "Data Transmission"
    pub. McGraw-Hill, Berkeley, CA., 1982

The Waite Group, "PC-LAN Primer"
    pub. Howard W. Sams & Co. Inc, Indianapolis, Indiana, 1986

"An Introduction to the IEEE 488 Bus Standard"
    pub. Farnell Instruments Ltd. Wetherby, UK. 1980

"Tutorial Description of the Hewlett-Packard Interface Bus"
    pub. Hewlett-Packard.

DEVICE INDEX

SUBJECT INDEX